Advances in Genetic Polymorphisms

Edited by Nouha Bouayed Abdelmoula and Balkiss Abdelmoula

Published in London, United Kingdom

Advances in Genetic Polymorphisms
http://dx.doi.org/10.5772/intechopen.1001553
Edited by Nouha Bouayed Abdelmoula and Balkiss Abdelmoula

Contributors
Prashant Vasisth, Omkar M. Limbalkar, Mohit Sharma, Hiral Sanghavi, Nouha Bouayed Abdelmoula, Samir Aloulou, Sonda Kammoun, Aymen Damak, Mona Rekik, Saloua Ben Amor, Oldez Kaabi, Balkiss Abdelmoula, Chrisanne Freeman, Julio Alejandro Valdez, Pedro Mayorga, Rafael Villa Angulo, Carlos Villa Angulo, Miruna Giurgiu, Robert Kaltenbach, Franziska Ahrend, Summer Weeks, Holly Clifton, Martin Bouldo, Vitaly Voloshin, Jiling Zhong, Siegfried Harden, Alexander Kofman

First published in London, United Kingdom, 2023 by IntechOpen
IntechOpen is the global imprint of INTECHOPEN LIMITED, registered in England and Wales, registration number: 11086078, 5 Princes Gate Court, London, SW7 2QJ, United Kingdom

British Library Cataloguing-in-Publication Data
A catalogue record for this book is available from the British Library

Additional hard and PDF copies can be obtained from orders@intechopen.com

Advances in Genetic Polymorphisms
Edited by Nouha Bouayed Abdelmoula and Balkiss Abdelmoula
p. cm.
Print ISBN 978-1-83769-637-6
Online ISBN 978-1-83769-636-9
eBook (PDF) ISBN 978-1-83769-638-3

Meet the editors

Nouha Bouayed Abdelmoula is a Professor and Senior Lecturer at the Faculty of Medicine, University of Sfax, Tunisia. She is the director of the Genomics of Signalopathies at the Service of Precision Medicine Laboratory (LR23ES07). She earned her MD, MSc, and Ph.D. degrees at the University of Tunisia and she received her training in Medical Genetics, Cytogenetics, Molecular Biology and Genetic Counseling at the Universities of France (Paris). She is a specialized physician (MD) in histology, embryology, and genetics and a pioneer researcher in biological engineering (Ph.D.) and molecular biology (MSc). Her research projects within LR23ES07 aim to promote translational medicine and expedite new paradigms in medicine and health while studying the signaling pathways in human development and diseases.

Balkiss Abdelmoula is a medical doctor specialized in family medicine and a public/global health expert. She earned her MD, MPH, and MSc in Data Analytics and Health Economics at the Medical University of Sfax, Tunisia. She has been a researcher at the Genomics of Signalopathies at the Service of Precision Medicine Laboratory (LR23ES07) since its establishment in 2017. She sculpted her career choices around serving human beings with a multidisciplinary approach, combining different disciplines in medicine. For more than a decade, Dr. Abdelmoula led several advocacy initiatives both nationally and internationally while holding leadership roles within different nongovernmental organizations. She is currently an associate member of the World Medical Association and a consultant at Global Health Strategies.

Contents

Preface

Single nucleotide polymorphisms (SNPs) are a prevalent form of genetic variation that contributes to the diversity among individuals. Initially, SNPs were utilized as biological markers to locate genes associated with diseases. Later, the identification and characterization of single nucleotide variations within the genome provided valuable insights into the genetic factors underlying human traits, diseases, and drug responses. Early methods, such as Sanger sequencing and restriction fragment length polymorphism (RFLP) analysis, though informative, had limitations in analyzing large-scale SNP datasets. However, with recent advancements in high-throughput genotyping technologies like microarrays and next-generation sequencing, SNP analysis has undergone a paradigm shift. These advanced platforms enable the simultaneous analysis of hundreds of thousands or even millions of SNPs across the nuclear and mitochondrial genomes and through the nucleic acids epigenome biomarkers, offering a comprehensive and cost-effective view of genetic variations. Technological advances have empowered scientists and clinicians to take a closer look at the molecular mechanisms underlying human health and diseases. This field has recently emerged as a feat advance in precision medicine because of its clinical relevance in diagnostic, prognostic, and predictive values. Furthermore, innovative algorithms and computational methods have emerged to effectively analyze and interpret the vast number of SNPs in large-scale SNP databases generated by high-throughput technologies through genome-wide association studies, a powerful tool for identifying statistically significant associations between specific SNPs and phenotypic outcomes of complex traits and diseases. Recently, these studies have paved the way for the exploration of the intricate interactions between genetic and epigenetic variations that govern gene expression regulation, leading to the gain of profound insights into the molecular mechanisms underlying the interplay between human health and the environment. In the realm of translational medicine, SNP analysis plays a crucial role in developing strategies for disease testing, prevention and treatment, investigating human evolution, predicting human traits, diseases, and responses to medications, and applying this knowledge to drug revelation and improvement and to personalized medicine for more accurate and efficient healthcare delivery. Spanning diverse applications from forensic analysis to pharmacogenomics and gene therapy, SNP analysis has become an indispensable tool for comprehending and enhancing human health.

Nouha Bouayed Abdelmoula and Balkiss Abdelmoula
Genomics of Signalopathies at the Service of Precision Medicine,
Medical University of Sfax,
Sfax, Tunisia

Section 1

SNP Genotyping Technological Progresses

Chapter 1

Genetics of Multiple Alleles: Concept and Function

Prashant Vasisth, Omkar M. Limbalkar and Mohit Sharma

Abstract

Mendelian genetics revealed only two alternative forms of a gene called alleles. The concept has evolved with the identification of more than two alternative forms of a gene, commonly referred to as multiple alleles. There are several traits that are governed by multiple alleles, such as ABO blood group system in humans, coat color in rabbits, and self-incompatibility in crop plants. The test of allelism is a very common practice to establish the relationship between alleles of the same or different genes. The inter-mating among different mutants helps to confirm whether mutations are allelic or non-allelic. The structural allelism determines whether two mutations are present at the same or different site in DNA and functional allelism determines whether two mutations are present in the same gene or in different genes. The concept of multiple alleles should not be confused with pseudoalleles and with pseudogenes. Pseudoalleles are two genetically linked genes with similar effects located close to each other on the chromosome, on the other hand, pseudogenes are nonfunctional copies of the functional genes. To understand the allelic relationships among and between genes is always a subject of interest. Therefore, in this chapter, the concept, function, and importance of multiple alleles are discussed.

Keywords: multiple alleles, mutation, pseudoalleles, pseudogene, allelism

1. Introduction

The extensive research of Gregor Johann Mendel on pea led to the discovery of the law of segregation and law of independent assortment, the two important principles of genetics. He studied seven traits with two contrasting phenotypes (wild and mutant) and indicated the presence of only two alleles for each gene [1]. Allele is referred to as an alternative form of a gene and locus is the location of allele in the genome or on chromosome in an organism. The concept of two alleles for each gene has changed with the discovery of more than two alleles for a gene as there is no restriction on number of alleles in a population. The presence of more than two alleles in a group of individuals is designated as multiple alleles (also referred to as allelic series). The concept of multiple alleles holds true for a population and it should not be misinterpreted at the individual level. A particular diploid organism can have at most two alleles from different alleles of the same gene on each homologous chromosome. However, many alternative forms of a gene can exist between different members of species. Multiple alleles have a similar inheritance pattern as that of two alleles but multiple alleles may have a large number of genotypes and phenotypes. The general formula to calculate the possible number of genotypes is n(n + 1)/2, where n stands for number of alleles involved. The possible number of homozygotes and heterozygotes are n and n(n-1)/2, respectively (**Table 1**).

Let us have a hypothetical example to understand the concept of multiple alleles at the molecular level. With understanding that an organism has a pair of homologous chromosomes each from male and female parent, and genes are present on the chromosomes and their positions on the chromosomes are referred to as locus/loci. Gene, on the other hand, is a sequence of nucleotides performing specific functions. The mutation causes a change in DNA sequence of a gene, leading to the emergence of an alternative form of a gene (allele/mutant). Multiple alleles, are thus, changed DNA sequences of a gene, arises due to mutation at a locus. In the following example, wild-type allele *A* has mutated to form three different alleles, that is, *a1*, *a2*, and *a3* (**Figure 1**).

Total number of alleles	Number of genotypes	Number of homozygotes	Number of heterozygotes
1	1	1	0
2	3	2	1
3	6	3	3
n	**n(n + 1)/2**	**n**	**n(n-1)/2**

Table 1.
Possible number of genotypes of multiple alleles.

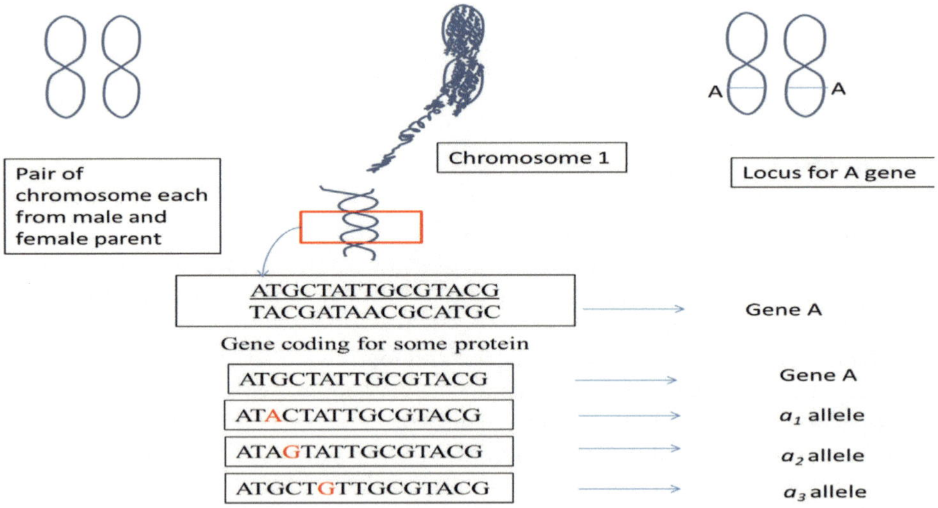

Figure 1.
Concept of gene, allele, and locus.

2. Symbols for mutant alleles

William Bateson designated dominant and recessive alleles of a particular gene with a single letter. He assigned the symbol *T* as the dominant allele for the tall pea plant and *t* as a recessive allele for the dwarf pea plant. Later, it became common to use gene symbols based on the mutant alleles/abnormal alleles. Afterward, two or more letters were used to represent different genes because a number of genes identified were greater than the number of English alphabet. To date, only two alleles of a gene were symbolized using uppercase and lowercase letters. Since a gene can have multiple alleles and symbolizing these multiple alleles has become more complicated. Therefore, a combination of gene symbols and superscripts were used together to symbolize multiple alleles, for example, alleles of a white locus in Drosophila. At this locus, pure white eye color is designated as w and other alleles as a superscript on w. Yellowish orange eye color was designated by w^{a} and w^{bf} was used for light buff eye color. Hyphenated symbols were also used to identify mutant allele for shrunken maize kernel represented as sh_2-6801. Further need was felt to assign specific symbols for differentiation of wild type from mutant-type allele. Few researchers used + sign as a superscript to the allele symbol or + sign alone for designating wild-type allele. Replacement of the first lowercase letter of the symbol of mutant-type allele with an uppercase letter to differentiate both. For example, Sh_2 was used for defining wild-type allele and sh_2 for mutant-type allele.

3. Important examples of multiple alleles

There are several examples of multiple alleles, such as blood group in human, coat color in rabbit, and self-incompatibility in plant, which helps in understanding this concept with ease.

3.1 ABO blood group system

Study on human blood type led to the discovery of ABO blood group system by Karl Landsteiner in the early 1900s [2]. He awarded with Nobel Prize in Physiology or medicine for the discovery of the human blood group in 1930. Serological studies were used to identify the A, B, and O antigens in which blood samples were tested with the different types of sera. Out of different sera, one serum specifically detects A antigen, and another serum detects B antigen. The presence of only "A" antigen is responsible for "A" blood group, "B" antigen is responsible for B blood type, AB blood type contains both A and B antigen and O blood type neither have A nor B antigen. A single gene with multiple alleles I^A, I^B, and i is responsible for the ABO blood group system where the symbol "I" stands for the isoagglutinogen or antigen. I^A allele is responsible for the synthesis of "A" antigen, I^B allele for B antigen, and i for the synthesis of O/H antigen [3]. Possible numbers of genotypes from three alleles are I^AI^A, I^Ai, I^BI^B, I^Bi, I^AI^B, and ii. These genotypes result into four distinguishable phenotypes as both I^AI^A, I^Ai *are* responsible for A type of blood group, I^BI^B, I^Bi will represent B type blood group, I^AI^B will represent AB type blood group and ii will responsible for O type of blood group. These phenotypes indicate the dominance relationships between three alleles. I^A and I^B alleles are dominant to i allele when either of them are present with i allele. When I^A and I^B alleles are present together, they alleles express equally and, thus, express codominance.

Biosynthesis of ABO blood group system required H substance precursor made up of galactose (Gal), N-acetylglucosamine (AcGluNH) [4]. ABO genes encode different types of glycotransferase enzymes. The function of glycotransferase enzymes is to add sugar groups to the preexisting polysaccharides. These polysaccharides with added sugar combined with lipids to form glycolipids and the association of red blood cells with glycolipids led to the formation of blood group antigens.

3.1.1 Synthesis of H/O antigen

Almost all individuals having glycolipids are called as H antigen. It is made up of chemically linked galactose (Gal), N-acetylglucosamine (AcGluNH), and Fucose. Addition of Fucose to the precursor was done by α 1,2 fucosyltranferase (H transferase) enzyme led to the formation of H antigen [5]. As it is commonly found in all individuals, one, two, or more different types of sugar groups can be added to it for the synthesis of other antigens. As *i* allele is recessive to both I^A and I^B allele, it is incapable of producing functional O transferse enzyme. Inactive enzyme cannot add any terminal sugar to the H antigen, therefore, only H antigen is present in O type of blood group (**Figure 2**).

3.1.2 Synthesis of "A" antigen

I^A allele is responsible for the synthesis of "A" antigen. I^A allele encodes acetylgalactosaminyltransferase (A transferase) enzyme, which is involved in the conversion of H antigen to "A" antigen. A transferase enzyme addition of N galactosamine to the H antigen with the help of A transferase enzyme led to the synthesis of "A" antigen.

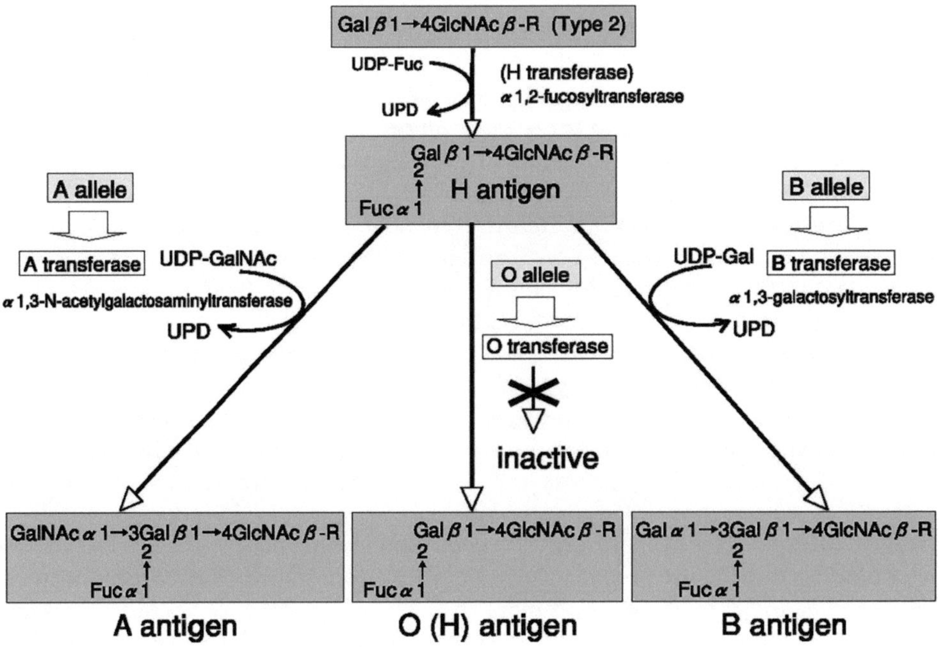

Figure 2.
Flow chart represents the biosynthesis of A, B, and O antigens [6].

3.1.3 Synthesis of "B" antigen

I^B allele encodes α 1,3 galactosyltransferase (B transferase) enzyme which is different from the enzyme produced by the I^A allele. B transferase enzyme is incapable of adding N galactosamine to the H antigen instead, it adds galactose to the H antigen.

Due to the difference in the DNA sequence of I^A and I^B alleles, they encode functionally different but highly related glycosyltransferase enzymes. The formation of AB blood group type is possible only when I^A and I^B alleles are present in a heterozygote state. As both alleles act co-dominantly, both enzymes are produced and a person with AB blood group type will be having both A and B antigens [7]. Some of the H antigens will be converted to the A antigens by the activity of A transferase and some of the H antigens will be converted to B antigens by the activity of B transferase.

The inheritance of a blood group is used to solve cases of disputed maternity or paternity. For example, parents with O blood type cannot have a child with AB blood type (alone these data are insufficient for legal purposes). Based on blood group inheritance, we cannot prove that an individual is a parent instead, we can only determine that an individual is not a parent of a particular child. Antigen-antibody relationship in the ABO blood group system is an important aspect in determining the protocol for blood transfusion [8].

Person with blood group A having genotypes as I^AI^A or I^Ai, produces only antigen A. Thus, people with blood type A possess antibodies against B antigen (Anti-B) as B antigen is a foreign molecule for them. If blood transfusion from a person with B blood type to one with A blood type, there will be agglutination (clump) of red blood cell due to the antigen-antibody interaction. Clumped cells are not able to move through the fine capillaries, leading to the failure of an organ or even death. An individual with blood group B having genotypes I^BI^B or I^Bi, produces only antigen B and antibodies against A. Individuals with AB blood type having genotype I^AI^B, produce both A and B antigens. These individuals neither have antibodies against A or B. In the case of O blood type having genotype ii, devoid of both antigens A and B. The blood serum of individuals with O blood type contains antibodies against antigens A and B (**Table 2**).

Individuals with A blood type can donate blood to people with blood type A or AB as they do not have antibodies against "A" antigen. The person with B blood type can donate blood to the ones with blood type B or AB as they do not have antibodies against B. Person with AB blood type can only donate blood to the ones with only blood type of AB because of the presence of both A and B antigen [9]. However, they can accept blood from any of the other blood types as they do not have antibodies against A and B antigens. Therefore, individuals with AB blood type are referred to as universal recipients. Individuals with O blood type can donate blood to a person with any of the other blood types because they are devoid of A and B antigens in their blood serum. But they can receive blood only from the people with O blood type only as both antigen A and B acts as foreign molecule in the blood serum of O blood type. Therefore, individuals with O blood type are regarded as universal donors (**Table 2**).

3.2 Bombay blood type

The H antigen is an important precursor for the ABO blood group system. It is produced by the activity of the α 1,2 fucosyltransferase (H transferase) enzyme, which is encoded by the dominant H gene. The H gene is distinct from the ABO blood group gene [10]. In rare cases, individuals may inherit two copies of a rare

Blood group phenotype	Genotype/alleles	Antigen present	Antibody produced	Special properties
A	I^AI^A, I^Ai	A antigen	B antibody	
B	I^BI^B, I^Bi	B antigen	A antibody	
AB	I^AI^B	A and B antigen	No antibodies	Universal recipient
O	*ii*	—	Both A and B antibody	Universal donor

Table 2.
ABO blood type system.

mutation in the dominant H gene, resulting in the inability to produce the H transferase enzyme, which leads to the absence of H antigen [11]. This is known as the Bombay phenotype or the h/h genotype. Individuals with the Bombay phenotype cannot produce A or B antigens, regardless of the presence of IA and IB alleles [12]. The absence of H antigen means that IA and IB alleles cannot recognize H substance, leading to the lack of A and B antigen synthesis. Individuals with the Bombay phenotype are similar to the O blood group type, but they are different in that they produce anti-O antibodies, whereas individuals with O blood type do not [13]. This information is useful in solving cases of disputed maternity or paternity, as it provides insight into the likelihood of biological relationships between individuals based on their blood group inheritance [14] (**Figure 3**).

3.3 Coat color in rabbits

Coat color in rabbits is governed by color determining gene (*c*) is another example of multiple alleles. Four alleles designated as c^+ for wild-type allele, c^{ch} for chinchilla-type allele, c^h for Himalayan-type allele, and *c* for albino-type allele determine the coat color [15]. These alleles developed by the mutation in the wild-type allele in the course of evolution [16]. The effect of allele can be seen when they are present in homozygous conditions. The majority of the rabbits are homozygous for the c^+, referred to as wild-type allele, in the wild population. Dominance relationships among multiple alleles of color determining genes in rabbits can be studied by making different heterozygous combination at this locus. Wild-type allele is completely dominant to the all-other alleles. Chinchilla allele is partially dominant in the Himalayan and albino alleles. The Himalayan allele is completely dominant in the albino allele.

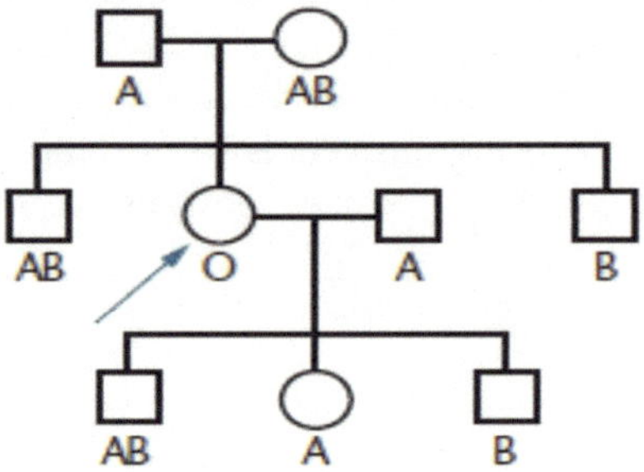

Figure 3.
Pedigree showing Bombay blood type [14].

DOI: http://dx.doi.org/10.5772/intechopen.1001464

Female parent (Stigma)	Male parent (Pollen)		
	S_1S_2	S_2S_3	S_3S_4
S_1S_2	Self-sterile	S_2S_3, S_3S_1	S_3S_1, S_3S_2, S_4S_1, S_4S_2
S_2S_3	S_1S_2, S_1S_3	Self-sterile	S_4S_2, S_4S_3
S_3S_4	S_1S_3, S_1S_4, S_2S_3, S_2S_4	S_2S_3, S_2S_4	Self-sterile

Table 3.
Different cross combinations of self-incompatibility in plant.

3.4 Self-incompatibility in plant

The self-incompatibility in the plant has been reported to be governed by multiple alleles [17]. Self-sterility or self-incompatibility is referred to as pollen of the plant unable to germinate on its own stigma and no fertilization in ovules takes place [18]. East and Mangelsdorf [19] observed multiple alleles are responsible for self-incompatibility in *Nicotiana*. The genes for self-incompatibility have been designated as S with allelic series S_1, S_2, S_3, S_4, and S_5.

The pollen produced on a plant unable to fertilize the ovule of the same plant and cross combination is completely self-sterile, such as S_1S_2 xS_1S_2, S_2S_3x S_2S_3, and S_3S_4 x S_3S_4 (**Table 3**). In the cross combinations, such as S_1S_2xS_2S_3, S_2S_3xS_1S_2, S_2S_3 xS_3S_4, and S_3S_4xS_2S_3, one of the alleles is common in both the parents and, therefore, one pollen is ineffective while other produce zygote. Similarly, in the cross S_1S_2xS_3S_4, S_3S_4xS_1S_2 all the pollens are effective and four kinds of progenies viz., S_1S_3, S_1S_4, S_2S_3, and S_2S_4 will be produced (**Table 3**).

4. Test of allelism

The alternate form of a gene is called as allele/mutant allele. The change in the base pair of the DNA sequence of a gene lead to the development of mutant allele. Mutant allele may or may not have predictable phenotype always. If a trait is governed by several genes and mutation occurs in one of the genes governing the trait, there will be a reduction or abolishment of phenotype. Prediction of a gene that has mutated on the basis of phenotype remains a challenge. Therefore, a test has to be conducted to predict which gene has mutated, provided mutation is recessive in nature. For this, an unknown recessive mutation is crossed with other recessive mutations of known genes. If we get wild-type phenotype as a product of crossing over, it means two mutations used in crossing were non-allelic in nature and if we get mutant type, then two mutations used in crossing were allele of the same gene [20]. It explains the principle that mutation for the same gene affects the same genetic function (**Table 4**).

Let us have an example of eye color mutants of fruit fly, *Drosophila melanogaster*, to understand the test of allelism by using three recessive eye color mutations. These mutations are *scarlet*, *cinnabar*, and *cinnabar-2* and we cannot differentiate these mutations on the basis of phenotype because all three mutations resulted into a bright red eye color phenotype, whereas, the wild type produces dark red eye color [21]. The test of allelism need to be performed for confirming whether mutations are alleles of single color determining gene or a different one. For this cross between these recessive homozygous strains of mutations can be made and a hybrid phenotype can be

Unknown recessive mutation	Known tester recessive mutation	Phenotype of hybrid (F_1)	Remarks
aa	bb	wild	a and b are non-allelic
	cc	mutant	a and c are allelic
	dd	wild	a and d are non-allelic

Table 4.
General scheme for the test of allelism.

Crosses	Phenotype of hybrid	Remarks
scarlet × *cinnabar*	Wild type (dark eye color)	*scarlet* and *cinnabar* are non-allelic
cinnabar × *cinnabar-2*	Mutant type (bright eye color)	*cinnabar* and *cinnabar-2* are allelic
scarlet × *cinnabar-2*	Wild type (dark eye color)	*scarlet* and *cinnabar-2* are non-allelic

Table 5.
A test of allelism involving recessive eye color mutations in Drosophila.

analyzed. If phenotype of the hybrid is wild type, then mutations used in the crosses are non-allelic. On the other hand, if the hybrid exhibit a mutant phenotype, then the mutation used in crosses will be allelic in nature. The phenotype of the hybrid between *scarlet* and *cinnabar* mutants was wild type, which indicates that these two mutations are in the different genes and are non-allelic. The hybrid phenotype obtained from the cross between *cinnabar* and *cinnabar-2* was found to be a mutant type, which means these two mutations are allelic (**Table 5**).

5. Structural and functional basis of allelism

Generally, the allelic relationship between different alleles can be established on two major aspects, such as structure and function, and it is named as structural allelism and functional allelism, respectively [22]. In the case of structural allelism, the occurrence of two or more mutations (governing the same trait) on the same site of the nucleotide sequence in DNA and is based on a recombination test whether two mutations can recombine or not at the DNA level. If the two mutations can recombine with each other through crossing over and producing wild-type allele, then they are structurally non-allelic, whereas, if two mutations do not recombine and are unable to produce wild-type DNA sequence, they are known to be structurally allelic in nature. Mutations on different sites can undergo the process of recombination, referred to as structurally non-allelic and mutations on the same site in DNA cannot recombine, referred to as structurally allelic (**Figure 4**). Functional allelism determines whether two mutations, governing the same trait, are present in the same gene or in the two different genes [23]. It is done by a complementation test (complementation at product level). If two individuals have mutations in the same gene, they are referred to as functionally allelic and wild-type phenotype will not develop after intermating of these mutants while if two individuals have mutations in two different genes, then wild-type phenotype will be produced through complementation [24], are referred as functionally non-allelic (**Table 6**, **Figure 4**). Therefore, if two mutations are structurally and

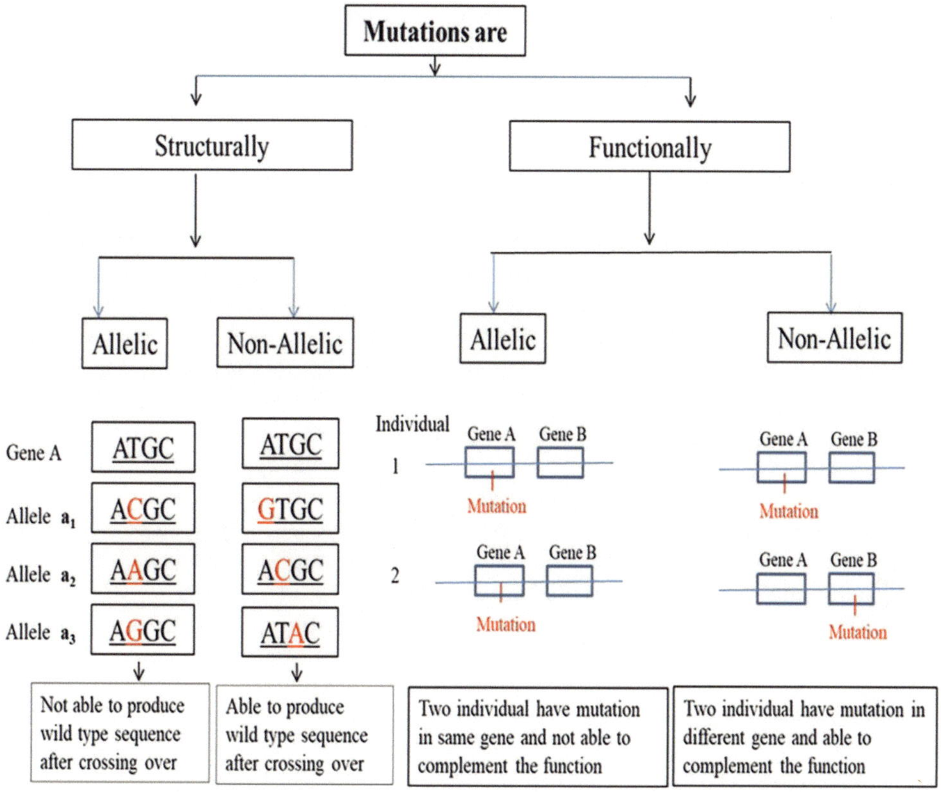

Figure 4.
Allelic relationship of different mutations.

Trait	Structural allelism	Functional allelism
Function	It determines location of two mutation in DNA sequences, help to understand whether two mutations can recombine to produce wild types	It determines whether two mutations present in the same gene or in two different genes and help to understand their complementation
Detection	Based on the recombination test	Based on the complementation test
Occurs at	DNA level	Product level
Breakage of chromosome	Yes	No

Table 6.
Distinguishing features of structural and functional allelism.

functionally allelic, they are called as homoalleles and if two mutations are functionally allelic but structurally non-allelic are referred to as heteroalleles.

6. Pseudoalleles vs. multiple alleles

Pseudoalleles are defined as two genetically linked genes with similar effects located close to each other on the chromosome [25, 26]. As two genes (pseudoalleles) are genetically linked, genes always tend to inherit together and may appear to act as a single gene [27]. Pseudoalleles should not be confused with pseudogenes, which are nonfunctional copies of functional genes, and arises directly through duplication or indirectly by reverse transcription of mRNA transcripts, while pseudoalleles are linked loci. The pseudoalleles differed from multiple alleles that former is linked loci situated close to each other while later is alternative form of single gene [28]. To distinguish pseudoalleles from multiple alleles, few shreds of evidence were shown in past such as (1) many crosses have been made between the mutants available at that time to resolve allelic series by crossing over resulted into failure (2) heterozygotes for the two different mutant genes have phenotypes intermediate of two respective homozygotes, but expected phenotype was non-allelic/wild type. Some cases have been reported where test for allelism comes positive for non-allelic genes due to the position effect. This phenomenon has been considered as positional pseudoallelism, as two genes occupy separate loci along with the position effect [29]. In positional pseudoallelism, the coupling phase heterozygote gives a wild type or nearly wild type (++/ab) phenotype and the repulsion phase (+a/b+) heterozygote gives mutant phenotypes.

Red eye color mutants of *Drosophila* such as white eye (*w*) and apricot (*apr*) eye color mutants with chromosome order as *apr-w* is an example of pseudoalleles/positional pseudoallelism [25]. Lewis [30] studied 21 crossovers between *apr* and *w* and found 0.03% of crossover between them with a map distance of 0.005 to 0.02 units. Due to the position effects, *apr+/+w* female gives mutant phenotype and *++/aprw* female gives wild-type phenotypes.

Pseudoalleles originated from the process of gene duplication. Two copies of the gene have been created by gene duplication, which remains closely associated on the chromosome but progressively diverges in structure and function [25]. Some of the distinguishing features of multiple alleles, pseudoallele, and pseudogene have been given in **Table 7**.

Trait	Pseudoallele	Pseudogene	Multiple allele
Definition	Pseudoalleles are defined as two genetically linked genes with similar effects located close to each other on the chromosome	Pseudogenes are non-functional copies of functional genes	The presence of more than two alleles in a group of individuals, designated as multiple alleles
Origin	By gene duplication	Duplication or indirectly by reverse transcription of mRNA	By mutation
Crossing over	Possible	Possible	Not possible
Final product	Functional	Non-functional	Functional
Locus involved	More than one locus	More than one locus	Only one locus
Trait affected	Same trait	Same trait	Same trait
Example	Red eye colour of Drosophila has different mutants like white and apricot	*PTENP1* pseudogene of tumor-suppressing gene.	ABO blood group system in humans

Table 7.
Distinguishing features of multiple alleles, pseudoallele, and pseudogene.

7. Conclusion

In this text, we have discussed the concept, function, and importance of multiple alleles. Multiple alleles arise due to mutations and result in different forms of a gene that can affect the traits of an organism. The study of allelic relationships among genes is essential for understanding genetic diversity and evolution. We have also discussed the differences between multiple alleles, pseudoalleles, and pseudogenes, and the methods used to distinguish them. Furthermore, we have explored the phenomenon of positional pseudoallelism and its effects on genetic inheritance. In conclusion, multiple alleles are a significant source of genetic variation that plays a crucial role in the evolution and genetic diversity. Understanding the allelic relationships between genes is essential for studying the genetics of organisms and for developing methods for genetic manipulation and disease prevention.

Author details

Prashant Vasisth*, Omkar M. Limbalkar and Mohit Sharma
Division of Genetics, ICAR-Indian Agricultural Research Institute, New Delhi, India

*Address all correspondence to: pvasisth555777@gmail.com

References

[1] Dunn LC. A Short History of Genetics. New York: McGraw-Hill; 1966

[2] Landsteiner K. Zur Kenntnis der antifermentativen, lytischen und agglutinierenden Wirkungen des Blutserums und der Lymphe. Zentralblatt Bak-teriologie. 1900;**27**:357-362

[3] Yoshida A. Biochemical genetics of the human blood group ABO system. American Journal of Human Genetics. 1982;**34**:1-14

[4] Ginsburg V. Enzymatic basis for blood groups. Methods Enzymology. 1972;**36**:131-149

[5] Misevic G. ABO blood group system. Asia-Pacific Journal of Blood Types and Genes. 2018;**2**(2):71-84

[6] Hosoi E. Biological and clinical aspects of ABO blood group system. Journal of Medical Investigation. 2008;**55**:174-182

[7] Peters JA. Classic Papers in Genetics. Englewood Cliffs, NJ, London: Prentice-Hall Inc; 1959

[8] Race RR, Sanger R. Blood Groups in Man. 6th ed. Oxford: Blackwell; 1975

[9] Landsteiner K, Levine P. Further observations on individual differences of human blood. Proceedings of the Society for Experimental Biology and Medicine. 1927;**24**:941-942

[10] Yunis EJ, Svardal JM, Bridges RA. Genetics of the Bombay phenotype. Blood. 1969;**33**(1):124-132

[11] Kelly RJ, Ernst LK, Larsen RD, Bryant JG, Robinson JS, Lowe JB. Molecular basis for H blood group deficiency in Bombay (Oh) and Para-Bombay individuals. Proceedings of the National Academy of Sciences. 1994;**91**(13):5843-5847

[12] Dipta TF, Hossain AZ. The Bombay blood group: Are we out of risk? Mymensingh Medical Journal: MMJ. 2011;**20**(3):536-540

[13] Bhar S, De A, Saha A, Bhattacharyya C. Pediatric patient with Bombay blood group: A rare case report. Saudi Journal of Anaesthesia. 2015;**9**(3):318

[14] Klug WS, Cummings MR, Spencer CA, Palladino MA, Ward SM. Concepts of Genetics. 10th ed. Vol. 9. USA: Pearson Education, Inc; 2009

[15] Punnett RC. Inheritance of coat-colour in rabbits. Journal of Genetics. 1912;**2**(3):221-238

[16] Foster M. Mammalian pigment genetics. Advance. Genetics. 1965;**13**:311-339

[17] Charlesworth D. Self-incompatibility. F1000 Biology Reports. 2010;**2**:68

[18] Castric V, Vekemans X. Invited review: Plant self-incompatibility in natural populations: A critical assessment of recent theoretical and empirical advances. Molecular Ecology. 2004;**13**(10):2873-2889

[19] East EM, Mangelsdorf AJ. A new interpretation of the hereditary behavior of self-sterile plants. Proceedings of the National Academy of Sciences of the United States of America. 1925;**11**(2):166

[20] Demerec M, Ozeki H. Tests for allelism among auxotrophs of Salmonella typhimurium. Genetics. 1959;**44**(2):269

[21] Ephrussi B, Beadle GW. Development of eye colors in Drosophila: Production and release of CN^+ substance by the eyes of different eye color mutants. Genetics. 1937;**22**(5):479

[22] Laughnan JR. Structural and functional bases for the action of the A alleles in maize. The American Naturalist. 1955;**89**(845):91-103

[23] Girard LR, Fiedler TJ, Harris TW, Carvalho F, Antoshechkin I, Han M, et al. WormBook: The online review of *Caenorhabditis elegans* biology. Nucleic Acids Research. 2007;**35**:D472-D475

[24] Rédei GP. Allelism Test. In: Encyclopedia of Genetics, Genomics, Proteomics, and Informatics. Dordrecht, Netherlands: Springer Science & Business Media; 2008. DOI: 10.1007/978-1-4020-6754-9_513

[25] Morange M. Pseudoalleles and gene complexes. Perspectives in Biology and Medicine. 2015;**58**(2):196-204

[26] Welshons WJ, Von Halle ES. Pseudoallelism at the notch locus in drosophila. Genetics. 1962;**47**(6):743

[27] Lewis EB. Pseudoallelism and gene evolution. In: Cold Spring Harbor Symposia on Quantitative Biology. New York, USA: Cold Spring Harbor Laboratory Press; 1951;**16**:159-174. DOI: 10.1101/sqb.1951.016.01.014

[28] Carlson EA. Allelism, complementation, and pseudoallelism at the dumpy locus in Drosophila melanogaster. Genetics. 1959;**44**:347

[29] Lewis EB. Some aspects of position pseudoallelism. The American Naturalist. 1955;**89**(845):73-89

[30] Lewis EB. The Pseudoallelism of white and apricot in *Drosophila melanogaster*. Proceedings of the National Academy of Sciences of the United States of America. 1952;**38**(11):953-961

Chapter 2

Recent Advancements in SNP Typing Methods Used in Forensic Science

Hiral Sanghavi

Abstract

Single nucleotide polymorphisms (SNPs) are heritable variations at defined regions and occur in at least 1% of the population. SNPs are mostly bi-allelic, and their inheritance pattern can be typed in a representative population of few unrelated individuals. Contrary to the STRs (Short tandem repeats), SNPs can be typed relatively easily using next generation sequencing methods. Thus, SNPs have attracted a lot of scientists for application in forensic analysis for cases such as establishing biogeographical ancestry, evolutionary timescale relatedness, immediate family relationships. SNPs are arguably more useful than STRs in certain forensic scenarios. For instance, when the obtained DNA sample from crime scene is a multi-origin mixture or when the DNA is degraded, SNPs offer better utility. SNPs are also valuable in cases where DNA extraction from challenging forensic samples, such as bones or meat, poses technical difficulties. Due to their characteristics, SNPs provide enhanced capabilities for forensic analysis in these specific situation. A plethora of novel techniques and algorithms have been developed to use the available SNP databases for forensic analysis. The developed technologies include hybridization assay, primer extension assay, multiplex polymerase chain reaction (PCR), denaturing high performance liquid chromatography, matrix-assisted laser desorption/ionization. These are just a few examples of the developed technologies utilized in molecular biology and genetic analysis, each with its unique advantages and disadvantages. We discuss the accuracy, sensitivity, specificity, advantages and disadvantages of some of these techniques in detail here.

Keywords: SNP, forensic, SNP typing, molecular beacons, oligonucleotide ligation assay

1. Introduction

Single nucleotide polymorphisms (SNPs) refer to genomic positions at which different nucleotides can occur in different individuals [1–3]. Any nucleotide variation at a genomic position can be considered as an SNP, only if the same variation occurs at a frequency of 1%, in a sampled population. SNPs are found abundantly in the human genome, on an average of 1 in every 1000 nucleotide [1–3]. Multiple databases have been developed over years to keep up with the ever-increasing number of reported SNPs. Some of these databases are: The SNP consortium (1998), dbSNP [4, 5], 1000 genomes project (2008) [6], the GWAS catalog (2008) [7], ALlele FREquency Database (ALFRED) [8]. Interestingly, these databases also store information on SNPs constituting the same haplotype, SNPs related to distinct disorders and distinct phenotypes.

Mutations often lead to formation of SNPs. The mutation can either be transition mutations (for example: A to G or G to A) or transversion mutations (for example: C to A or G to C). Transition mutation led SNPs account for bi-allelic loci and transversion mutation led SNPs account for bi-allelic, tri-allelic or tetra-allelic loci (**Figure 1**). Bi-allelic SNPs are found more abundantly in the human genome [1, 9, 10]. It is interesting to note that a few thousands of the reported SNPs existed in the human genome before humans migrated out of South Africa. These SNPs account for the common SNPs found across different human populations on the earth. In addition to these common SNPs, there are thousands of SNPs reported as population-specific SNPs which may have arisen due to distinct biogeographical location, epigenetic factors such as food, lifestyle preferences, local environmental conditions. SNPs have been used for a range of purposes, including human identity verification, differentiation amongst distinct biogeographic human populations, analysis of human immigration patterns, exploration of ancestral history, determination of sibling relationships and identification of criminals across the border [11–13].

SNPs provide a genetic tool to be used in defining human populations, distinguishing one individual from the other. However, Short Tandem Repeats (STRs) have been a conventional choice for similar applications. STRs are short DNA sequences (about 2–6 nucleotide long) which make up about 3% of the total human genome [14–16]. Lately, SNPs have gained popularity as an alternative to STRs due to characteristic properties such as: bi-allelic nature of SNPs, low mutation rates as compared to STRs, inheritance of SNPs in the form of haplotypes, relation to distinct phenotypes, easy identification from smaller amplicons (often obtained from degraded forensic samples), lack of stutter artifacts, amenable to multiplexing assay and automation [1–3]. However, the use of SNPs has faced certain limitations. These include the need for a large number of SNPs compared to STR markers to achieve discrimination between human and non-human samples. Careful selection of population-specific SNPs is necessary, which can be challenging. SNPs have limited utility in mixture interpretations due to their bi-allelic nature. Additionally, available criminal offender databases such as The Combined DNA Index System (CODIS) [17] predominantly contain STR typing datasets, necessitating the

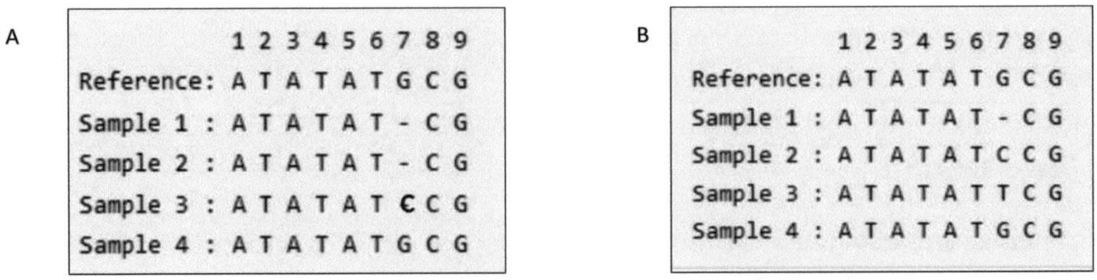

Figure 1.
Single nucleotide polymorphisms (SNPs) can be bi-allelic, tri-allelic or tetra-allelic. (A) Schematic representation of a bi-allelic SNP, presence of C in sample 3 and G at position 7 in sample 4. (B) Schematic representation of a bi-allelic SNP, presence of C in sample 2, T at position 7 in sample 3 and G at position 7 in sample 4.

retyping of these databases for SNPs. Developing multiplexing assays for SNPs can be costly and time-consuming, and the resulting information may be less informative compared to STRs.

Multiple SNP typing methods have been developed for their use in basic research and forensic applications. With the advent of technological advancement, many of these methods can be adapted for multiplexing assays. SNP typing methods are based on four different reaction principles: (1) Hybridization, (2) Oligonucleotide ligation, (3) Primer extension and (4) Enzymatic cleavage [2, 3]. The reaction principles have been combined with different combinations of assay formats and detection platforms for use of SNPs in user-specific requirements. Below we describe various methods based on reaction principles, advancements in the methods, and combinations of methods with different assay formats and detection platforms.

2. SNP typing methods based on the principle of hybridization

2.1 Allele-specific hybridization

DNA–DNA hybridization (DDH) involves identification of a target DNA using a complementary DNA probe (short strand of DNA) [18]. DDH has been conventionally used to differentiate between alleles using allele-specific DNA probes [19]. Allele-specific hybridization attempts to identify DNA targets at a polymorphic locus using Allele-Specific Oligonucleotide probes (ASO). The probes are designed such that a central polymorphic nucleotide is flanked by gene specific-DNA sequence. The complementarity between target DNA and the probe determines the stability of hybridization [20]. Two probes, one complementary to each allele, are used in each hybridization reaction. One of the two probes will be completely complementary to the target DNA, thus forming a stable hybrid (**Figure 2**). Optimized reaction

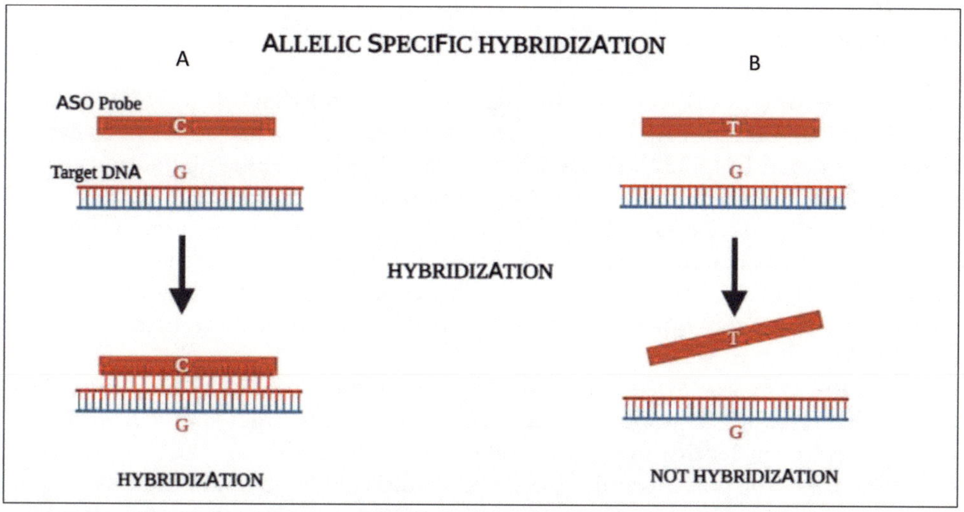

Figure 2.
Allele specific hybridization. The allele specific hybridization depends on the design of probes. Allele specific oligonucleotide probes (ASO) have a polymorphic site at the Centre of the probe. (A) Schematic representation of hybridization of ASO to the target DNA (B) schematic representation of no hybridization between ASO and target DNA due to mismatch at the polymorphic site.

conditions favor the formation of stable hybrid. Detection of the stable hybrid has been attempted with various techniques. However, fluorescence resonance energy transfer (FRET) has remained the detection technique of choice in combination with allele-specific hybridization.

2.2 Fluorescence resonance energy transfer (FRET)

FRET works on the principle of energy transfer from a donor fluorophore to an acceptor fluorophore. Excitation of the donor fluorophore at a wavelength (within its excitation spectrum) leads to emission of energy. This emitted energy serves as a source of excitation of the acceptor fluorophore. The energy transfer is possible because the emission spectrum of the donor overlaps with the excitation spectrum of the acceptor. FRET is a distance sensitive technique, i.e., the donor fluorophore and the acceptor fluorophores have to be in close proximity for a successful energy transfer [21].

2.3 Recent advancements in the method

ASO and FRET were restricted by the amount of DNA available for studies. Retrieval of very small quantities of DNA is often faced by forensic scientists. This affects the sensitivity, efficiency and the accuracy of SNP typing. To overcome this, ASO and FRET were combined with polymerase chain reaction (PCR). PCR uses a target DNA and amplifies it using primers (short DNA sequences; complementary to target DNA), DiDeoxy nucleotide tri phosphates (dNTPs), DNA polymerase. Primers provide the 3′ OH group, as a substrate for addition of dNTPs complementary to target DNA, by DNA polymerase. Below we described the three technical advancements depending on the design of the probe, using exonuclease activity of the DNA polymerase. Each of these three advanced methods attempts to enhance sensitivity, accuracy and efficiency of the conventional hybridization method.

2.4 LightCycler®

LightCycler® (*Roche*) was one of the earlier platforms that offered SNP typing based on ASO, PCR and FRET. It used two probes that hybridize adjacent to each other on the target DNA [22]. Probe 1 had a central polymorphic nucleotide and was labeled with a fluorophore at its 3′ end and probe 2 was labeled with a different fluorophore at its 5′ end. FRET signal was generated only if the two probes had hybridized adjacent to each other because of the distance sensitivity of FRET (**Figure 3A**). The FRET signal was detected and quantified by the LightCycler® instruments. The signal quantity is directly proportional to the amplification of target DNA.

Earlier, the assay was designed to use different probe sequences for typing different SNPs. The size of the probes may be similar for distinct target DNAs, however, the polymorphic nucleotide in the probe would vary depending on the target DNA. The difference in the polymorphic nucleotide would lead to different melting temperature of the probe. In case of unknown target DNA, melting curve analysis post PCR reactions determined the SNP in the target DNA. With time, LightCycler® has been updated for use in real-time quantification (LightCycler® 480 Instrument II), diagnostics by multiplexing and endpoint genotyping applications (LightCycler® 96 Instrument) [23].

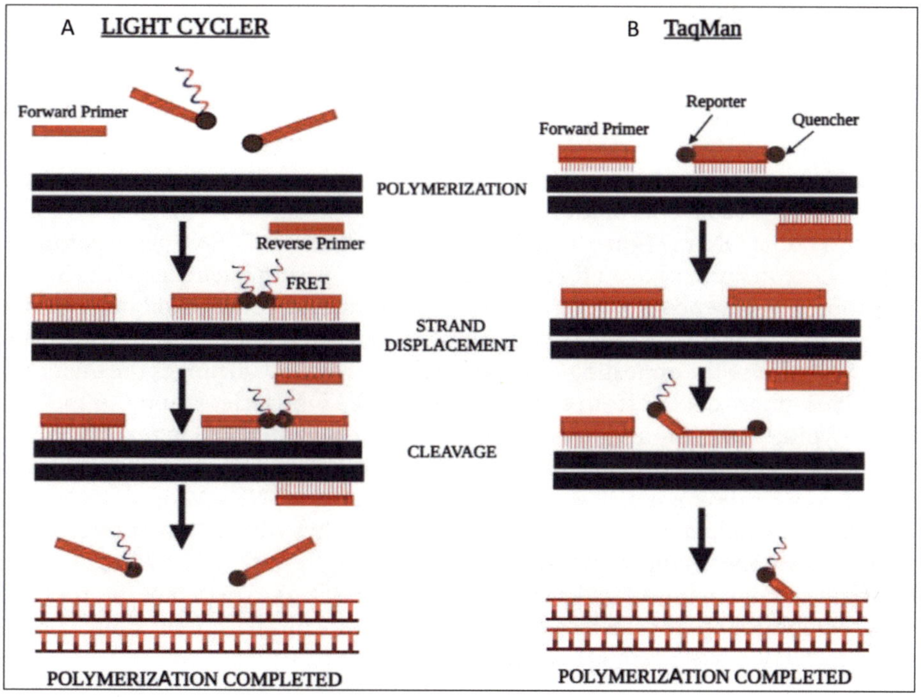

Figure 3.
Recent methods of allele specific hybridization which use FRET for detection. (A) Light cycler uses two ASO, each labeled with a fluorophore at its end. Generation of FRET signal determines the adjacent hybridization of both ASO to the target DNA. (B) TaqMan uses a single ASO labeled with a fluorophore and a quencher. FRET signal determines cleavage of ASO from its 5′ end by Taq polymerase.

2.5 TaqMan™ SNP genotyping assay

TaqMan™ SNP genotyping assay (*ThermoFisher Scientific*) was a contemporary of the LightCycler®. TaqMan™ SNP genotyping assay takes advantage of the 5′ exonuclease activity of the Taq DNA polymerase, used in the PCR reactions [24–26]. Like the LightCycler®, TaqMan™ SNP genotyping assay combined ASO, PCR and FRET. However, it required only one probe per reaction as opposed to two probes per reaction in the LightCycler®. The probe was attached to a fluorophore at the 5′ end and attached to a quencher at the 3′ end. Close vicinity of the fluorophore and the quencher does not allow any fluorescence emission when the probe is intact. Probes anneal to the target DNA during annealing step in the PCR. During the extension step, the Taq DNA polymerase extends the primer sequence and cleaves the probe from 5′ to 3′ as it comes in the way of DNA synthesis. Cleavage is a result of the 5′ exonuclease activity of the Taq DNA Polymerase. The emission of fluorescence signal from the probe is a measure of whether the probe had annealed to the target DNA (**Figure 3B**). Like the LightCycler®, the intensity of the fluorescent signal signifies the amount of target DNA multiplied [2, 3]. Conventional TaqMan™ assay utilized two probes. Each probe designed complementary to different polymorphic sites or alleles. Different fluorophores were attached to different probes. Thus, the FRET emission would be used to correlate with the sequence of the target DNA [24–26]. Recently TaqMan™ assay has been adapted for use in multiplex reactions using different combinations of fluorophores and quenchers [27].

2.6 Molecular beacons

Molecular Beacons mark the latest advancements in ASO probes. The probes are designed such that they have DNA sequence complementary to the target DNA as well as DNA sequence with intra-molecular complementarity. The intra-molecular complementarity is the characteristic feature of the molecular beacon probes. Intra-molecular complementarity leads to formation of a hair pin loop structure of the probe until it anneals to a target DNA (**Figure 4**) [28–30]. Like the TaqMan™ SNP genotyping assay, molecular beacon probes have a fluorophore at the 5′ end and quencher at the 3′ end of the probe. The absence of fluorescence signal signifies that probe did not anneal to the target DNA. The fluorescence signal signifies that the probe has annealed to the target DNA because the annealing of probe to the target DNA arranges the quencher and the fluorophore distant from each other. Like the LightCycler® and the TaqMan™ SNP genotyping assay, the amplification of DNA can be monitored in real time using the intensity of the fluorescent signal [1, 2]. Conventionally, two different molecular beacon probes are designed for two different alleles and/or SNPs. Multiplexing reactions have been conducted using molecular beacon probes with different fluorophores for each allele or SNP in the target DNA and a wavelength shifting molecular beacons. Wavelength shifting molecular beacons are probes attached to different fluorophores. Each of these fluorophores can be excited using a monochromatic light source or a laser but each fluorophore emits a distinct wavelength (distinct color) [31].

2.6.1 Array hybridization

All three of the above technologies had restricted multiplexing capacity due to: availability of distinct, non-overlapping fluorophores emitting in selected wavelength spectrum, similar reaction conditions (annealing temperatures, melting temperatures, product lengths) for multiple target DNAs to be amplified simultaneously [1, 2]. This was overcome upon selecting a microarray assay format to screen multiple SNPs simultaneously. Microarrays have multiple probes attached to the solid surface. These probes are used to hybridize with different fluorescently labeled PCR products. Each PCR product represents a distinct SNP (**Figure 5**). However, the hybridization depends on the length of the sequences flanking the polymorphic site [32]. GeneChip® system

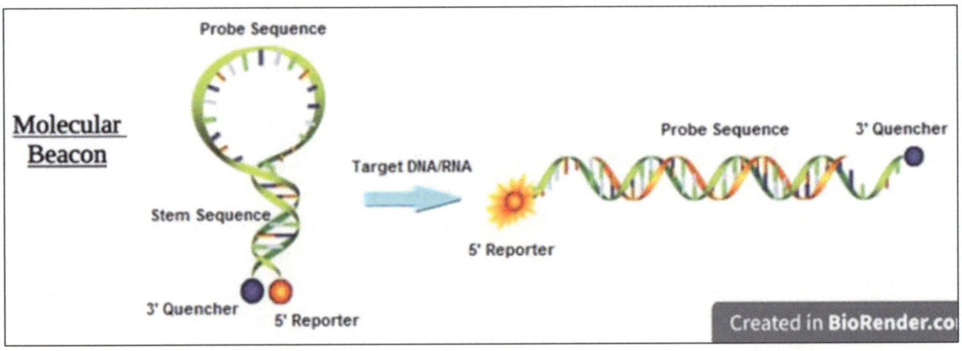

Figure 4.
Design of molecular beacon probes used for advanced allele specific hybridization. The molecular beacon probe has intra-molecular complementary sequence leading to formation of a hair pin structure, flanked by a fluorophore at the 5′ end and quencher at the 3′ end. Upon hybridization to the target DNA, the fluorophore and the quencher are placed distant to each other emitting fluorescence signal.

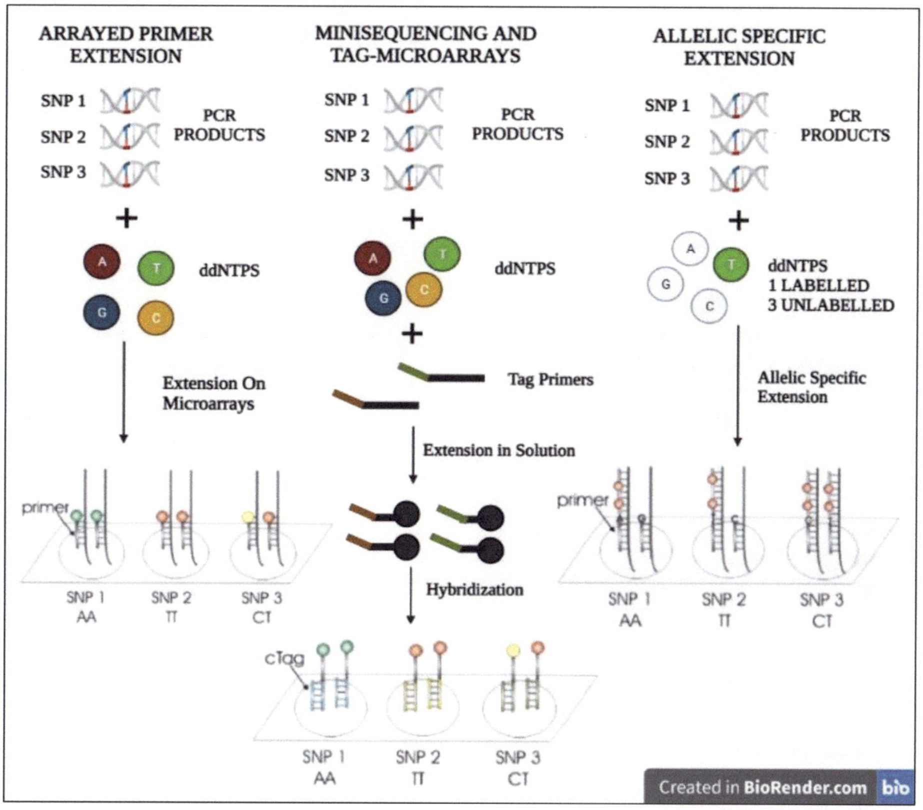

Figure 5.
Array hybridization. A microarray format for typing multiple SNPs by attaching multiple probes to the solid surface. Probes are used to hybridize with the PCR products. Each PCR product will represent a distinct SNP. Using an array of probes aids identification of many SNPs simultaneously.

used multiple ASO probes for the same SNP, spanning all the possible combinations of the polymorphic site and the flanking sequence (Tiling strategy) [33–35]. Tiling strategy could overcome the ambiguity in the microarray results arising due to the difference in lengths of the sequences flanking the polymorphic site. The multiplexing capacity of the GeneChip® system is beyond the actual need of SNP typing in forensic studies lately. However, it may prove to be useful for future applications.

3. SNP typing methods based on the principle of primer extension

Primer extension works on the principle of addition of nucleotides complementary to the target DNA using 3′ end of the primer as a starting point. The ability of DNA polymerase to accurately add nucleotide at the 3′ end of a primer sequence forms the basis of the primer extension reaction (**Figure 6**) [36]. The choice of length extended DNA, use of allele-specific primers has offered adaptations of primer extension reaction. These are: (A) Minisequencing, (B) Pyrosequencing and (C) Allele-specific primer extension [1, 2]. Irrespective of the length of extended DNA or distinct primers, the extended DNA was conventionally separated using electrophoresis technique for a length-based identification of polymorphic sites in the target DNA.

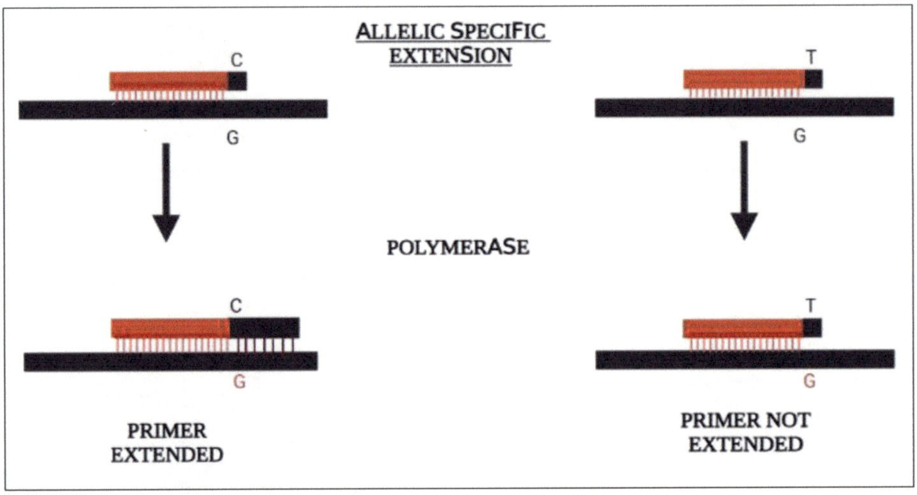

Figure 6.
Primer extension method of SNP typing. A recent method of SNP typing using principle of primer extension. Primers are designed to be allele specific such that primer extension happens only if the polymorphic site on the primer and the target DNA are complementary.

However, with technological advancements, fluorescence-based detection and mass spectrometry-based detection have been combined with primer extension reactions for accurate polymorphism identification.

3.1 Minisequencing

It refers to (i) sequencing few nucleotides using labeled DiDeoxy Nucleotide Tri Phosphates (ddNTPs) to detect polymorphic sites (**Figure 6**) or (ii) identifying the sequence of the target DNA using polymorphic site-specific primers [37, 38]. Case (i) is a conventional PCR except for the use of ddNTPs instead of dNTPs. Primers are designed such that they anneal the target DNA till immediately upstream of the polymorphic nucleotide (**Figure 7**). The addition of first ddNTP to the 3′ end of the primer identified the sequence at the polymorphic site. Labelling each ddNTP with distinct fluorophores provides real-time detection of the polymorphic nucleotide identification using fluorescence detectors, thus facilitating automation of the entire process. Extended DNA length may overlap between two independent primer extension reactions. Electrophoretic separation of overlapping lengths of DNA proved a challenge for accurate identification of polymorphic nucleotide. Addition of non-human DNA sequences at the 5′ end of the primers solved the problem of overlapping DNA lengths [1, 2]. SNaPshot™ was one of the earliest SNP typing methods that combined minisequencing with detecting a fluorescently labeled ddNTP [39, 40]. Currently the commercially available SNaPshot™ Multiplex Kit offers multiplexing of ten SNPs [41]. These have been used for forensic investigations and applications. Case (ii) is a conventional PCR as well. However, the primers are designed complementary to the polymorphic site of the target DNA. The primer extension happened only if the primer annealed to the target DNA. Thus, the polymorphic site can be identified by the primer sequence of the extended DNA. Use of mass spectrometry for detection of polymorphic site along with primer extension reaction has been used lately [42]. Mass spectrometer

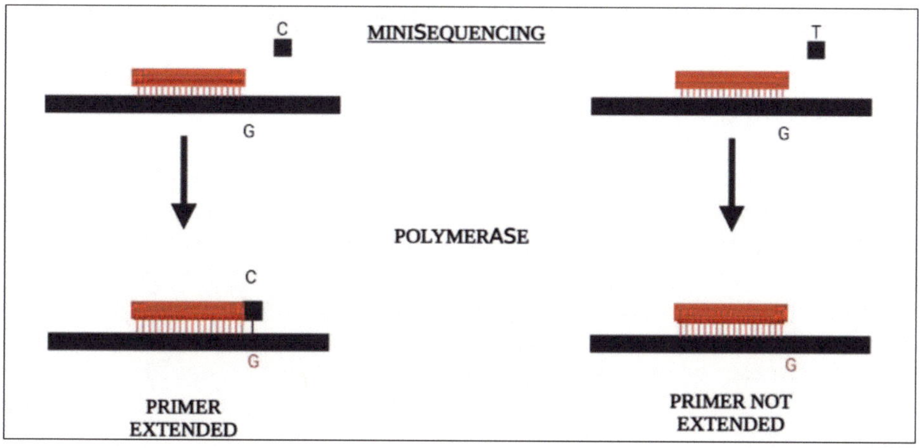

Figure 7.
Minisequencing method of SNP typing. SNP typing using sequencing the polymorphic nucleotide immediately downstream of the primer using distinctly labeled ddNTPs. The label on the annealed ddNTP complementary to the target will determine its sequence.

identifies the polymorphic site by differentiation in the mass of extended DNA. The difference in the mass of extended DNA is contributed by distinct mass of each ddNTP. Matrix Assisted Laser Desorption Ionization – Time of Flight (MALDI-TOF) is a preferred ionization source and detector combination used for SNP typing of minisequenced products [42]. PROBE assay [43], PinPoint assay [44], GOOD assay [45] are few of the SNP typing assays that use the minisequencing reaction principle coupled with mass spectrometry detection platform. Each of these assays attempts to distinguish extended DNA more accurately than the other. The PROBE assay utilizes both ddNTPs and dNTPs to increase the mass difference between the extended DNA. The PinPoint assay utilizes dNTPs only. The GOOD assay uses modified primer that enhances the purification of the primer extended DNA for detection using mass spectrometry. Like the SNaPshot™ assay, non-overlapping masses of extended DNA are achieved upon addition of non-human DNA sequences at the 5′ end of the primers.

3.2 Pyrosequencing

It was developed as a method to sequence longer DNA as compared to minisequencing. It works on the principle of sequencing by synthesis. Like a conventional PCR, the reaction mixture contains dNTPs, DNA polymerase, primers. In addition to these, the reaction mixture also contains ATP sulfurylase, luciferase, apyrase. Upon addition of every nucleotide, light is released. The light released is due to the pyrophosphate released upon activity of DNA polymerase. This pyrophosphate acts as a substrate of ATP sulfurylase, which converts it into ATP. Thus formed ATP is used by luciferase to generate light. Excess nucleotides are degraded by apyrase. Nucleotides are added to the sequencing reaction mixture in a pre-defined order. The release of light signifies incorporation of the added nucleotide. Thus, the sequence of the synthesized strand is determined based on the pre-defined order of addition of nucleotides. About 20–30 nucleotides can be sequenced using pyrosequencing [46, 47]. The advantage of this is being able to read the sequences around the polymorphic sites in the target DNA. However, limited multiplexing possibilities and lengthy target preparation procedures make pyrosequencing applicable to limited studies [1, 2].

3.3 Allele-specific primer extension

It involves the use of allele-specific primers and the target DNA. Only those primers which anneal stably (without-mismatch) to the target DNA will be extended by the DNA polymerase whereas the primers with polymorphic site other than the target DNA (mismatched primers) do not get extended [48, 49]. The detection platforms coupled with allele-specific primer extension vary from the conventional electrophoresis, fluorescence signal detection. Multiplexing with allele-specific primer extension has been possible by using microarray technology. Microarray allows detection of multiple polymorphic sites simultaneously or use of FRET primer pairs for real-time SNP typing (**Figure 4**) [50].

4. SNP typing methods based on the principle of oligonucleotide ligation

4.1 Oligonucleotide ligation assay (OLA)

DNA ligases are a class of enzymes which catalyze the formation of a phosphodiester bond between two subsequently placed nucleotides. The reaction which uses DNA ligase for sealing a nick between two consecutive nucleotides is known as oligonucleotide ligation. OLA uses DNA ligase to ligate two probes positioned one phosphodiester bond apart on a target DNA [51]. Of the two required probes, one probe is designed complementary to the polymorphic site on the target DNA at its 3′ end. The second probe, also known as the common probe, is complementary to the target DNA sequence immediately downstream of the polymorphic site. Ligation of the two probes happens only if both the probes have annealed to their respective target DNA (**Figure 8**). Thus, upon ligation, the probe sequence can be used to identify the polymorphic nucleotide in the target DNA [1, 2, 51].

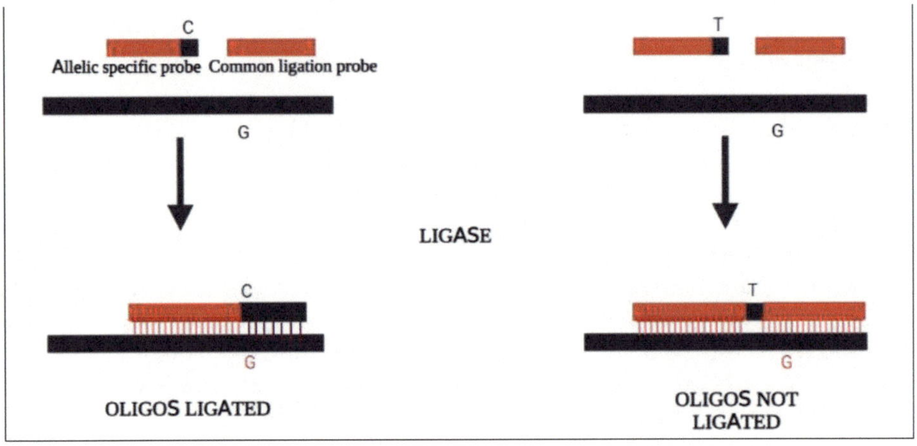

Figure 8.
Oligonucleotide ligation assay (OLA) for SNP typing. The OLA uses two probes. Both of them hybridize with the target DNA adjacent to each other. One of the probes are designed specific to the allele in the target DNA. Ligation of the two probes happens only if the probe hybridizes completely to the target DNA.

4.2 Recent advancements in the method

Ligase chain reaction (LCR) combines the OLA and PCR. LCR is designed such that includes a thermostable DNA ligase as a part of the conventional PCR mixture [1, 2, 52]. The ligated probes in the early steps of the reaction form a target for attachment of probes in the next ligation reaction. Use of two strands of the target DNA, one representing each allele, in an LCR helps to identify allele-specific polymorphic sites. Although, it would need two distinct probes (in addition to the common probe), one complementary to each allele at the polymorphic site of the target DNA [1, 2, 52].

Detection of ligated probes in the OLA and LCR has been attempted using biotin-based detection method. The common probes are biotinylated, and the allele-specific probes have a reporter dye attached to them. Ligation of the two probes will lead to capture of the ligated product using streptavidin [1, 2, 52]. Advancements in the detection technologies have replaced the biotin with mobility modifiers and replaced reporter dyes with fluorescent dyes. Mobility modifiers are chemical molecules, often oligomeric which bind to the ligated probes and define its mobility. Thus, distinguishing the ligated probes from the unligated probes. Use of the mobility modifiers served for higher resolution separation of ligated products upon electrophoresis. Importantly, the mobility modifiers regulate the mobility of the ligated products independent of the length of the ligated product. Availability of varied fluorescent dyes provides the possibility of multiplexing the assay. Detection of multiple SNPs is possible after electrophoretic separation of the ligation products and detection of their respective reporter dyes [1, 2, 52].

Coupled Amplification and Oligonucleotide Ligation (CAL) offers a one-step amplification and ligation reaction [53]. Using primers with high melting temperatures and probes with lower temperatures was the key to this. The reagents for both DNA amplification and oligonucleotide ligation are added to the same tube. Firstly, the primers anneal to the target DNA and amplifies the target DNA. The annealing temperature should be higher than the melting temperature of the oligonucleotide probes. In the second step, the ligation happens at a lower temperature, allowing hybridization of the probes which is a pre-requisite of the OLA. Advanced CAL was developed with fluorescently labeled probes for real-time SNP typing. Currently, SNPlex (ThermoFisher Scientific) genotyping system is a commercially available SNP typing method based on OLA [54]. It is used to perform SNP typing taking genomic DNA as the target DNA. More importantly, at times it has been used with fragmented genomic DNA. Multiple unlabelled oligonucleotide probes are used for multiplexing the assay, followed by amplification of the identified target DNA using universal primers. Thus, the recognized polymorphic sites are read using ZipChute probes. These probes hybridize to the target DNA, they are purified and sequenced using capillary electrophoresis.

5. SNP typing methods based on the principle of enzymatic cleavage

5.1 Enzymatic cleavage

Structure-specific Flap endonucleases (FEN) are enzymes which recognize overlapping nucleotides (fork-like structure) and cleave the DNA which hangs loosely (not bound to any DNA strand) (**Figure 9**) [55, 56]. Archaeal FENs were

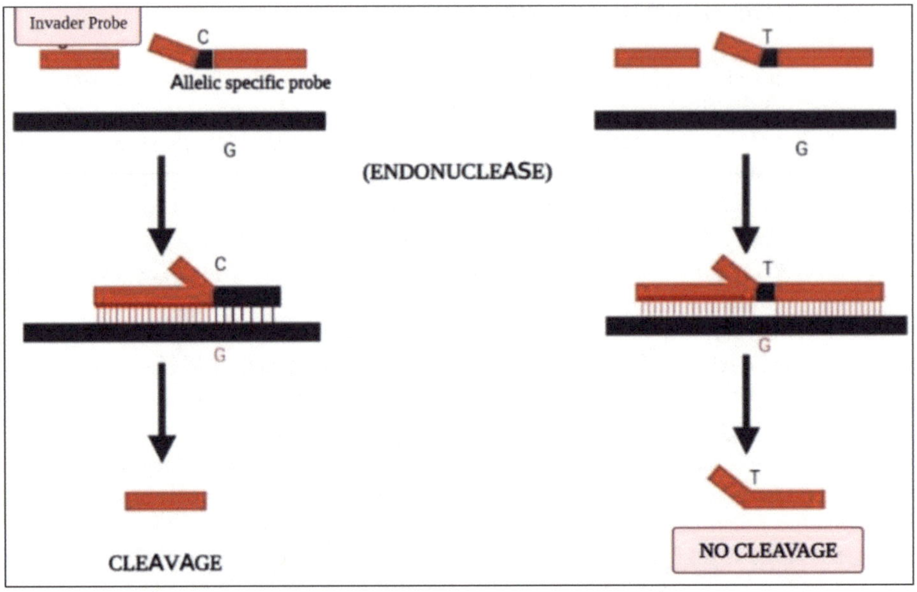

Figure 9.
Invader assay for SNP typing. The assay uses two probes. Allele specific probe and invader probe. Both of these probes have an overlap of one nucleotide. If the allele specific probe is complementary to the target DNA, a fork-like structure is formed between the invader probe and the allele specific probe. This structure is recognized by the FEN endonucleases mediating cleavage at the junction between probes.

earlier studied to recognize overlapping oligonucleotides which are hybridized to a target DNA strand and cleave the loosely hanging oligonucleotides. Interestingly, this cleavage was found to be sequence dependent as well [55, 56]. Thus, making the FENs a perfect fit for the enzymes which could be used for detection of polymorphic sites in a given target DNA using overlapping probes (complementary to the target DNA).

5.2 Invader assay

Invader assay was one of the first SNP typing methods used in forensic studies [57]. It works on the principle of cleavage of probe annealed to the target DNA at the polymorphic site. Each assay utilizes two probes: An allele-specific probe and an invader oligo. The invader oligo is designed such that it is complementary to the polymorphic site of the target DNA at the extreme 3′ end. The allele-specific probe is designed such that it is complementary to the polymorphic site on the target DNA towards it's 5′ end with some non-complementary nucleotides following the polymorphic site. The design of the probes leads to formation of an invader structure upon probes annealing to the target DNA. FENs recognize the invader structure which is like a bifurcated fork-like structure and cleave at the 3′ end of the overlapping invader structure (**Figure 9**). The cleavage can be detected in real time using FRET probes which have a fluorophore at the 5′ end of the probe and a quencher at the 3′ end of the probe [55–57]. Cleavage separates the quencher and the fluorophore generating a measurable fluorescent signal. In case of mismatch at the polymorphic site, the invader oligo will not anneal to the target DNA, hence no invader structure is formed, and no FEN cleavage is observed.

DOI: http://dx.doi.org/10.5772/intechopen.1001899

Despite the simple chemistry and one step method for SNP detection, the assay could not be adapted for forensic studies due to requirement of large amount of target DNA and minimized possibility of multiplexing the reactions [1, 2].

Acknowledgements

The authors thank Mr. Niranjan Vyas for helping create the figures used in this book chapter. The authors express heartfelt gratitude towards the School of forensic science, National Forensic Sciences University, for encouraging to initiate a book chapter on the recent advanced techniques and providing conducive environment for critical thinking and analysis of the subject matter.

Conflict of interest

The authors declare no conflict of interest.

Author details

Hiral Sanghavi
School of Forensic Sciences, National Forensic Sciences University, Gandhinagar, Gujarat, India

*Address all correspondence to: hiral.sanghavi@nfsu.ac.in

References

[1] Novroski NM, Cihlar JC. Evolution of single-nucleotide polymorphism use in forensic genetics. WIREs Forensic Science. 2022;**4**(6):e1459

[2] Sobrino B, Carracedo A. SNP typing in forensic genetics: A review. Methods in Molecular Biology. 2005;**297**:107-126. DOI: 10.1385/1-59259-867-6:107

[3] Syvänen AC. Accessing genetic variation: Genotyping single nucleotide polymorphisms. Nature Reviews. Genetics. 2001;**2**:930-942

[4] Smigielski EM, Sirotkin K, Ward M, Sherry ST. dbSNP: A database of single nucleotide polymorphisms. Nucleic Acids Research. 2000;**28**(1):352-355. DOI: 10.1093/nar/28.1.352

[5] Sherry ST, Ward M-H, Kholodov M, Baker J, Phan L, Smigielski EM, et al. dbSNP: The NCBI database of genetic variation. Nucleic Acids Research. Jan 2001;**29**(1):308-311

[6] Auton A, Brooks LD, Durbin RM, Garrison EP, Kang HM, Korbel JO, et al. A global reference for human genetic variation. Nature. 1 Oct 2015;**526**:68-74

[7] Buniello A, MacArthur JAL, Cerezo M, Harris LW, Hayhurst J, Malangone C, et al. The NHGRI-EBI GWAS Catalog of published genome-wide association studies, targeted arrays and summary statistics 2019. Nucleic Acids Research. 2019;**47**(D1):D1005-D1012. DOI: 10.1093/nar/gky1120

[8] Rajeevan H, Osier MV, Cheung KH, Deng H, Druskin L, Heinzen R, et al. ALFRED: The ALelle FREquency database. Update. Nucleic Acids Research. 2003;**31**(1):270-271. DOI: 10.1093/nar/gkg043

[9] Brookes AJ. The essence of SNPs. Gene. 1999;**234**(2):177-186. DOI: 10.1016/s0378-1119(99)00219-x

[10] Phillips C, Amigo J, Carracedo Á, Lareu MV. Tetra-allelic SNPs: Informative forensic markers compiled from public whole-genome sequence data. Forensic Science International. Genetics. 2015;**19**:100-106. DOI: 10.1016/j.fsigen.2015.06.011

[11] Mcevoy BP, Powell JE, Goddard M, Visscher PM. Human population dispersal "out of Africa" estimated from linkage disequilibrium and allele frequencies of SNPs. Genome Research. 2011;**21**(6):821-829

[12] Stoneking M, Krause J. Learning about human population history from ancient and modern genomes. Nature Reviews. Genetics. 2011;**12**(9):603-614. DOI: 10.1038/nrg3029

[13] Huisman J. Pedigree reconstruction from SNP data: Parentage assignment, sibship clustering and beyond. Molecular Ecology Resources. 2017;**17**(5):1009-1024. DOI: 10.1111/1755-0998.12665

[14] Wyner N, Barash M, McNevin D. Forensic autosomal short tandem repeats and their potential association with phenotype. Frontiers in Genetics. 2020;**11**:884. DOI: 10.3389/fgene.2020.00884

[15] Kayser M. Forensic use of Y-chromosome DNA: A general overview. Human Genetics. 2017;**136**(5):621-635. DOI: 10.1007/s00439-017-1776-9

[16] Haddrill PR. Developments in forensic DNA analysis. Emerging Topics in Life Sciences. 2021;**5**(3):381-393. DOI: 10.1042/ETLS20200304

[17] Bruce B, Moretti Tamyra R, Niezgoda Stephen J., Brown Barry L. CODIS and PCR-Based Short Tandem Repeat Loci: Law Enforcement Tools. Promega [Retrieved October 8, 2016]

[18] Kennell DE. Principles and practices of nucleic acid hybridization. Progress in Nucleic Acid Research and Molecular Biology. 1971;**11**:259-301. DOI: 10.1016/s0079-6603(08)60330-x

[19] Howell W, Jobs M, Gyllensten U, et al. Dynamic allele-specific hybridization. Nature Biotechnology. 1999;**17**:87-88

[20] Tortajada-Genaro LA. Design of Oligonucleotides for allele-specific amplification based on PCR and isothermal techniques. Methods in Molecular Biology. 2022;**2392**:35-51. DOI: 10.1007/978-1-0716-1799-1_3

[21] Kaur A, Kaur P, Ahuja S. Förster resonance energy transfer (FRET) and applications thereof. Analytical Methods. 2020;**12**(46):5532-5550. DOI: 10.1039/d0ay01961e Erratum in: Anal Methods. 2021;13(5):730

[22] Lareu M, Puente J, Sobrino B, Quintáns B, Brión M, Carracedo A. The use of the LightCycler for the detection of Y chromosome SNPs. Forensic Science International. 2001;**118**(2-3):163-168. DOI: 10.1016/s0379-0738(01)00386-3

[23] https://lifescience.roche.com/global/en/products/others/lightcycler-480-multiwell-plate-96-clear-381435.html

[24] Glaab WE, Skopek TR. A novel assay for allelic discrimination that combines the fluorogenic 5′ nuclease polymerase chain reaction (TaqMan) and mismatch amplification mutation assay. Mutation Research. 1999;**430**(1):1-12. DOI: 10.1016/s0027-5107(99)00147-5

[25] Holland PM, Abramson RD, Watson R, Gelfand DH. Detection of specific polymerase chain reaction product by utilizing the 5′----3′ exonuclease activity of Thermus aquaticus DNA polymerase. Proceedings of the National Academy of Sciences of the United States of America. 1991;**88**(16):7276-7280. DOI: 10.1073/pnas.88.16.7276

[26] Livak KJ, Flood SJ, Marmaro J, Giusti W, Deetz K. Oligonucleotides with fluorescent dyes at opposite ends provide a quenched probe system useful for detecting PCR product and nucleic acid hybridization. PCR Methods and Applications. 1995;**4**(6):357-362. DOI: 10.1101/gr.4.6.357

[27] https://www.thermofisher.com/in/en/home/life-science/pcr/real-time-pcr/real-time-pcr-assays/snp-genotyping-taqman-assays/taqman-snp-genotyping-assay-sets.html

[28] Tyagi S, Kramer FR. Molecular beacons: Probes that fluoresce upon hybridization. Nature Biotechnology. 1996;**14**(3):303-308. DOI: 10.1038/nbt0396-303

[29] Tyagi S, Bratu DP, Kramer FR. Multicolor molecular beacons for allele discrimination. Nature Biotechnology. 1998;**16**(1):49-53. DOI: 10.1038/nbt0198-49

[30] Kostrikis LG, Tyagi S, Mhlanga MM, Ho DD, Kramer FR. Spectral genotyping of human alleles. Science. 1998;**279**(5354):1228-1229. DOI: 10.1126/science.279.5354.1228

[31] Tyagi S, Marras SA, Kramer FR. Wavelength-shifting molecular beacons. Nature Biotechnology. 2000;**18**(11):1191-1196. DOI: 10.1038/81192

[32] Budowle B, Planz JV, Campbell RS, Eisenberg AJ. Single nucleotide

polymorphisms and microarray Technology in Forensic Genetics - development and application to mitochondrial DNA. Forensic Science Review. 2004;**16**(1):21-36

[33] Pease AC, Solas D, Sullivan EJ, Cronin MT, Holmes CP, Fodor SP. Light-generated oligonucleotide arrays for rapid DNA sequence analysis. Proceedings of the National Academy of Sciences of the United States of America. 1994;**91**(11):5022-5026. DOI: 10.1073/pnas.91.11.5022

[34] Dalma-Weiszhausz DD, Warrington J, Tanimoto EY, Miyada CG. The affymetrix GeneChip platform: An overview. Methods in Enzymology. 2006;**410**:3-28. DOI: 10.1016/S0076-6879(06)10001-4

[35] Wang DG, Fan JB, Siao CJ, Berno A, Young P, Sapolsky R, et al. Large-scale identification, mapping, and genotyping of single-nucleotide polymorphisms in the human genome. Science. 1998;**280**(5366):1077-1082. DOI: 10.1126/science.280.5366.1077

[36] Wu R. Development of the primer-extension approach: A key role in DNA sequencing. Trends in Biochemical Sciences. 1994;**19**(10):429-433. DOI: 10.1016/0968-0004(94)90094-9

[37] Syvänen AC. From gels to chips: "minisequencing" primer extension for analysis of point mutations and single nucleotide polymorphisms. Human Mutation. 1999;**13**(1):1-10. DOI: 10.1002/(SICI)1098-1004(1999)13:1

[38] Babol-Pokora K, Berent J. SNP-minisequencing as an excellent tool for analysing degraded DNA recovered from archival tissues. Acta Biochimica Polonica. 2008;**55**(4):815-819

[39] Geppert M, Roewer L. SNaPshot® minisequencing analysis of multiple ancestry-informative Y-SNPs using capillary electrophoresis. Methods in Molecular Biology. 2012;**830**:127-140. DOI: 10.1007/978-1-61779-461-2_9

[40] Schlebusch CM, Naidoo T, Soodyall H. SNaPshot minisequencing to resolve mitochondrial macro-haplogroups found in Africa. Electrophoresis. 2009;**30**(21):3657-3664. DOI: 10.1002/elps.200900197

[41] https://www.thermofisher.com/order/catalog/product/4323151

[42] Haff LA, Smirnov IP. Single-nucleotide polymorphism identification assays using a thermostable DNA polymerase and delayed extraction MALDI-TOF mass spectrometry. Genome Research. 1997;**7**(4):378-388. DOI: 10.1101/gr.7.4.378

[43] Braun A, Little DP, Köster H. Detecting CFTR gene mutations by using primer oligo base extension and mass spectrometry. Clinical Chemistry. 1997;**43**(7):1151-1158

[44] Paracchini S, Arredi B, Chalk R, Tyler-Smith C. Hierarchical high-throughput SNP genotyping of the human Y chromosome using MALDI-TOF mass spectrometry. Nucleic Acids Research. 2002;**30**(6):e27. DOI: 10.1093/nar/30.6.e27

[45] Sauer S, Lechner D, Berlin K, Lehrach H, Escary JL, Fox N, et al. A novel procedure for efficient genotyping of single nucleotide polymorphisms. Nucleic Acids Research. 2000;**28**(5):E13

[46] Lavebratt C, Sengul S, Jansson M, Schalling M. Pyrosequencing-based SNP allele frequency estimation in DNA pools. Human Mutation. 2004;**23**(1):92-97. DOI: 10.1002/humu.10292

[47] Ahmadian A, Gharizadeh B, Gustafsson AC, Sterky F, Nyrén P,

Uhlén M, et al. Single-nucleotide polymorphism analysis by pyrosequencing. Analytical Biochemistry. 2000;**280**(1):103-110. DOI: 10.1006/abio.2000.4493

[48] Russom A, Tooke N, Andersson H, Stemme G. Single nucleotide polymorphism analysis by allele-specific primer extension with real-time bioluminescence detection in a microfluidic device. Journal of Chromatography. A. 2003;**1014**(1-2):37-45. DOI: 10.1016/s0021-9673(03)01033-1

[49] Rejali NA, Moric E, Wittwer CT. The effect of single mismatches on primer extension. Clinical Chemistry. 2018;**64**(5):801-809. DOI: 10.1373/clinchem.2017.282285

[50] Vallone PM, Just RS, Coble MD, Butler JM, Parsons TJ. A multiplex allele-specific primer extension assay for forensically informative SNPs distributed throughout the mitochondrial genome. International Journal of Legal Medicine. 2004;**118**(3):147-157. DOI: 10.1007/s00414-004-0428-5

[51] Landegren U, Kaiser R, Sanders J, Hood L. A ligase-mediated gene detection technique. Science. 1988;**241**(4869):1077-1080. DOI: 10.1126/science.3413476

[52] Grossman PD, Bloch W, Brinson E, Chang CC, Eggerding FA, Fung S, et al. High-density multiplex detection of nucleic acid sequences: Oligonucleotide ligation assay and sequence-coded separation. Nucleic Acids Research. 1994;**22**(21):4527-4534

[53] Eggerding FA. A one-step coupled amplification and oligonucleotide ligation procedure for multiplex genetic typing. PCR Methods and Applications. 1995;**4**(6):337-345. DOI: 10.1101/gr.4.6.337 PMID: 7580927

[54] https://assets.thermofisher.com/TFS-Assets/LSG/Specification-Sheets/cms_059523.pdf

[55] Lyamichev V, Mast A, Hall J, et al. Polymorphism identification and quantitative detection of genomic DNA by invasive cleavage of oligonucleotide probes. Nature Biotechnology. 1999;**17**:292-296. DOI: 10.1038/7044

[56] Olivier M. The invader assay for SNP genotyping. Mutation Research. 2005;**573**(1-2):103-110. DOI: 10.1016/j.mrfmmm.2004.08.016

[57] Nakahara H, Sekiguchi K, Hosono N, Kubo M, Takahashi A, Nakamura Y, et al. Criterion values for multiplex SNP genotyping by the invader assay. Forensic Science International. Genetics. 2010;**4**(2):130-136. DOI: 10.1016/j.fsigen.2009.07.005

Section 2

SNP Genotyping Biomedical Applications Advances

Chapter 3

PCR Assay Using 22q11.2 Highly Polymorphic Markers for Exclusion of 22q Microdeletion: Technical Optimization and Application in North Africa

Nouha Bouayed Abdelmoula, Samir Aloulou, Sonda Kammoun, Aymen Damak, Mona Rekik, Saloua Ben Amor, Oldez Kaabi and Balkiss Abdelmoula

Abstract

22q11.2 deletion syndrome is a genomic disorder with a broader clinical and genetic spectrum. To exclude the presence of 22q11.2 microdeletion, we optimize a PCR-RFLP analysis of three SNP located in the typically proximal 22q11.21 deleted region of 1.5 Mb. PCR reactions, optimized with a Touch-Down program, were performed using three pairs of primers. The amplicons were cleaved by three restrictive enzymes: HaeIII, CviAII, and BsrI applied respectively, for rs4819523, rs4680, and rs5748411. The efficiency of this PCR RFLP assay was confirmed in the light of its application in a small cohort of 10 Tunisian patients, having a congenital heart defect and a known status of 22q11 deletion by FISH and MLPA. The principle of the proximal 22q11.2 microdeletion, applied with exclusion technique seems to be interesting but further population studies for the determination of the heterozygosity rate of the polymorphic 22q11 region markers are needed, particularly in North Africa.

Keywords: 22q11 microdeletion, karyotype, FISH, MLPA, north African population, PCR-RFLP, polymorphic markers, SNP

1. Introduction

22q11.2 deletion syndrome is a genomic disorder with a broader clinical and genetic spectrum [1–8]. Microalterations at the 22q11.2 locus include the most frequent 3 Mb deletion and the smallest nested 1.5 Mb deletion [2, 4, 5]. Non-allelic recombination events between LCRs A and H are responsible for 22q11.2 genomic rearrangements [6]. Molecular diagnosis of 22q11.2 microdeletion is usually made by fluorescence in situ hybridization (FISH), multiplex ligation-dependent probe

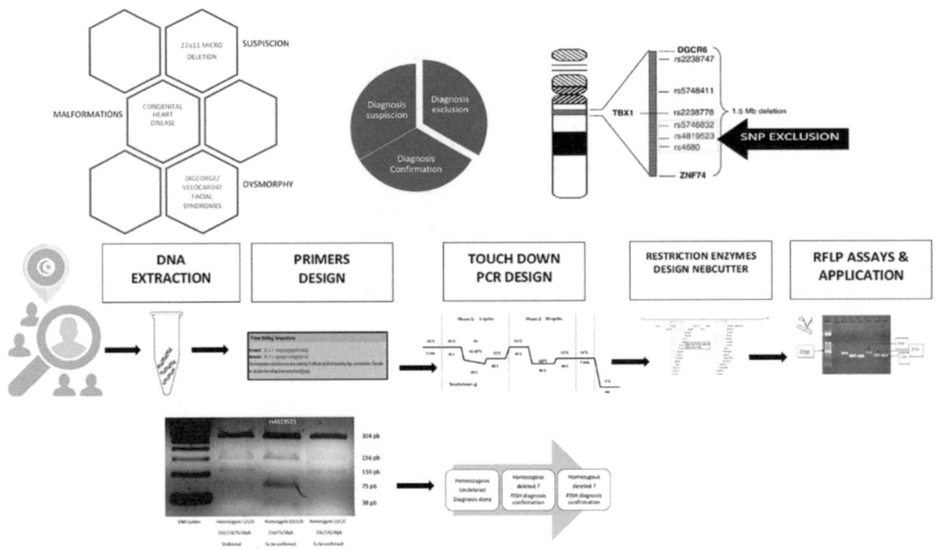

Figure 1.
Graphical illustration of our study design of 22q11 microdeletion exclusion.

amplification (MLPA), and comparative genomic hybridization (CGH) microarray analysis, or by real-time polymerase chain reaction (qPCR) [3].

Since 2003, it has been shown that a three-step PCR–RFLP analysis can be useful to exclude the presence of 22q11.2 microdeletion in the typically deleted 22q11 region of 1.5 Mb with less cost, high specificity, and with the benefit of reducing the number of the patients who are supposedly screened for 22q11.2 microdeletion (from 100 to 1%) [9]. This method was developed on the basis of the presence of polymorphism in consecutive selected SNP markers, located in the 1,5 Mb 22q11.2 microdeleted region, with a high heterozygosity rate in Brazilian population. Heterozygosity excluded 22q11.2 deletion. However, homozygosity is unable to exclude deletion and leads to a subsequent analysis of 22q11 region by FISH or MLPA [9–11].

Here, an efficient PCR-RFLP for 22q11.2 microdeletion exclusion was set up. It was evaluated using the technique of a small cohort of known deleted and undeleted Tunisian patients (**Figure 1**).

2. Material and methods

2.1 Patients

The application of the optimized PCR-RFLP technique was performed on 10 Tunisian patients having congenital heart diseases (CHD), and 22q11 known status. All patients were explored by karyotyping. The presence or the absence of 22q11.2 microdeletion was confirmed by FISH, using two LSI probes, which are respectively, LSI D22S75 (N25 region) SpectrumOrange/LSI ARSA SpectrumGreen Probe and DiGeorge Region Probe - LSI TUPLE 1 SpectrumOrange/LSI ARSA SpectrumGreen Probe. For two patients, the confirmation of 22q11.2 microdeletion was conducted by

MLPA, which was performed using the MRC-Holland SALSA MLPA P300 control kit, and designed synthetic probes targeting chromosomal regions, including the classical and distal 22q11.22 microdeletion region of DiGeorge/Velo-cardio-facial syndrome (Not shown).

2.2 Methods

2.2.1 Cytogenetic analysis

Cytogenetic analysis was carried out using lymphocyte cultures for 72h with phytohemagglutinin induction. Metaphase chromosomes were prepared in accordance with the standard cytogenetic protocols. Chromosome identification and karyotyping analysis were conducted by using RHG and GTG banding. Chromosomal abnormalities were interpreted and written in conformity with the international standard nomenclature (ISCN) 2005–2009.

2.2.2 Simplex PCR-RFLP: Technical optimization and analysis

A simplex PCR-RFLP was optimized for a rapid negative diagnosis of proximal microdeletions of chromosome 22, based on the method of molecular exclusion described by Pereira (2003) and Gioli-Pereira (2006) [9, 10].

To exclude the presence of 22q11.2 microdeletion, the method was conducted by the analysis of the polymorphism of the three loci located in the typically 22q11.21 deleted region of 1.5 Mb. These loci were rs4819523, rs5748411, and rs4680. They are characterized by a high rate of heterozygosity.

The interrogation of the SNP database (**Figure 2**) showed that these polymorphic markers included the ancestral alleles reported in some populations as the European, the Asian, and the Sub-Saharan African people.

The rs4819523 is a SNP (SNV) at the position: 19810549 with two alleles C/G (C is the ancestral one) (**Table 1**). It is an intronic variant of GNB1L gene (G-protein beta subunit-like protein ID 54584 location: 22q11.21 exon count: 8).

The rs4680 is a SNP (SNV) at the position: 1254184 with two alleles G/A (the G allele is the ancestral one) (**Table 1**) localized in COMT gene (Catechol-O-methyltransferase ID 1312 location 22q11.21 exon count: 8).

The rs5748411 is a SNP (SNV) at the position: 19733950 with two alleles C/A (C is the ancestral allele) (**Table 1**).

In addition, the interrogation of the SNP database showed that there is no information about the characteristics and the frequencies of the allelic polymorphisms of the three selected SNP (rs4819523, rs4680, and rs5748411) in the North African population.

PCR reactions were performed using three pairs of primers as described by Huber (2014) (**Table 2**) [11].

They were optimized by a Touch-Down program. Reactions were done, in a final volume of 25 µl, by using 10 pmol of each primer, 1x PCR buffer, 2 mmol/l of MgCl2, 0,2 mmol/l of dNTP, 200 ng DNA, 1 unit Taq DNA polymerase. Amplification conditions were optimized using 35 cycles Touch-Down program with initial melting step at 95°C for 5 min and final extension step at 72°C for 7 min in all reactions: five first cycles with 95°C for 40 s/ 65 to 60°C for 40 s and 72°C for 40 s and then 30 cycles with 95°C for 40 s/60°C for 40 s and 72°C for 40 s (**Table 3**).

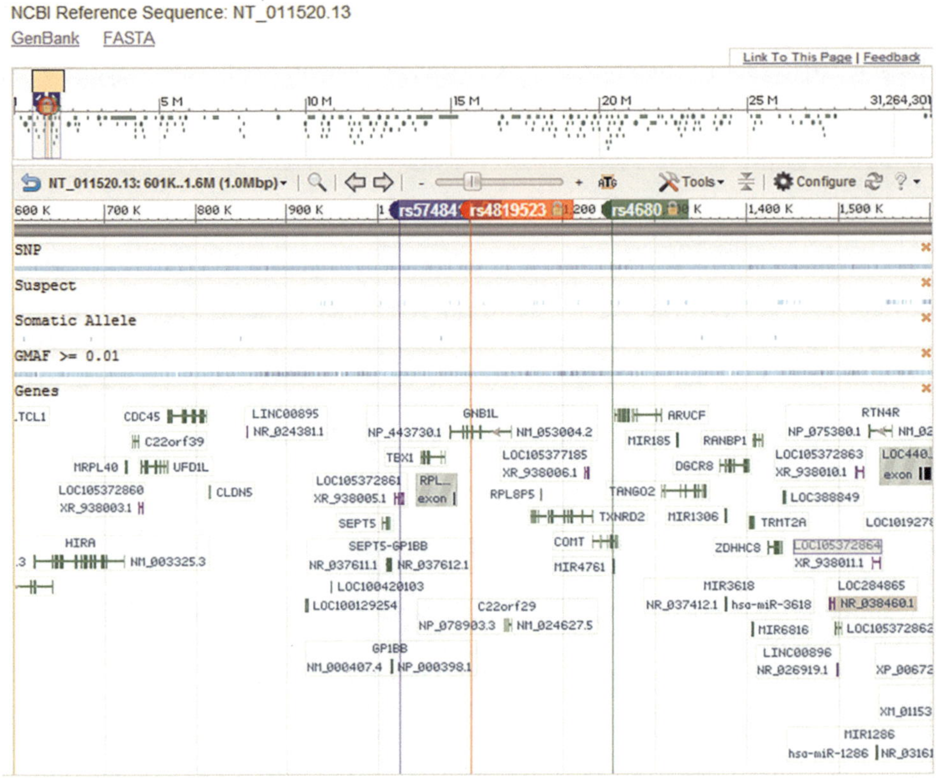

Figure 2.
Alignment of rs4819523, rs5748411, and rs4680 at the Homo sapiens chromosome 22 NCBI reference sequence: NT_011520.13 (https://www.ncbi.nlm.nih.gov/snp/).

Marker	Alleles and sequences of the ancestral alleles with the colored base in red
rs4819523	AGAGGTACACACCACGTCTACCTGG[C/G]CTGGGTCCTCAGCCCGGGGCTGCCC
rs4680	CCAGCGGATGGTGGATTTCGCTGGC[A/G]TGAAGGACAAGGTGTGCATGCCTGA
rs5748411	TTGCACCATGTGGCAGCTGTACCTG[A/C]CAGTCAGGGGATCCCAAGCTGGGGT

Table 1.
Sequences of ancestral alleles of the analyzed 22q11 polymorphic markers.

Marker	Forward PRIMERS	Reverse PRIMERS	Amplicon size	Alleles
rs4819523	ACAGTGAGGGGACATCAAGG	GGAAAGCTCATGGGTGTCTG	304 pb	C/G
rs4680	GTGATTCAGGAGCACCAGCC	GCCCTTTTTCCAGGTCTGAC	217 pb	G/A
rs5748411	ATCAGCCCACAGAGCAGACC	GCATAGAGAAAAGCAACTCC	202 pb	C/A

Table 2.
Sequences of primers used for PCR-RFLP validated by PCR in Silico.

For the RFLP analysis, the amplicons were cleaved by three restriction enzymes as described by Huber (2014) [11], in conformity with the manufacturer's recommendations (NEBiolabs) (**Table 4**).

Thermocycling Conditions			
Initial Denaturation	**95°C**	**5 min**	**1 cycle**
			5 cycles
Denaturation	95°C	40 s	
Annealing	65 °C → 60°C	40 s	
Extension	72°C	40s	
			30 cycles
Denaturation	95°C	40 s	
Annealing	60°C	40 s	
Extension	72°C	40 s	
Final extension	72°C	7 min	1 cycle
Hold	4%	∞	

Table 3.
The thermocycling conditions of the touch-down PCR reaction.

SNP	Restriction Enzyme	Restriction Sequence	Cleavage Site
rs4819523	HaeIII	GGCC	5′ ... GG↓CC ... 3′ 3’ ... CC↑GG ... 5′
rs4680	BclI	TGATCA	5′...T↓G A T C A...3′ 3′...A C T A G↑T...5’
rs5748411	BsrI	ACTGG	5′ ... ACTGGN↓ ... 3′ 3′ ... TGAC↑CN ... 5’

Table 4.
Restriction enzymes of RFLP and electrophorese profiles after cleavage.

The digestion of amplicons by the three restriction enzymes was optimized using three digestion programs (X1, X2, X3) and restriction fragments (**Table 5**) were visualized on 4% agarose gel. For X1 program: Reaction was performed, in a final

Marker	Restriction Enzyme	Cleavage fragments in the presence of the cleavage site	Cleavage fragments in the absence of the cleavage site
rs4819523	HaeIII	Allele C: 4 bands 156pb 75pb 38pb 35pb	Allele G: 3 bands 156pb 110pb 38pb
rs4680	BclI	Allele A: 2 bands 102pb 115pb	Allele G: 1 band 217pb
rs5748411	BsrI	Allele C: 2 bands 64pb 138pb	Allele A: 1 band 202pb

Table 5.
Electrophoretic profiles after marker digestion according to Huber et al. 2014.

volume of 20 µl, with 10 µl of PCR product, 2 µl of enzymatic tampon, 1 µl enzyme, and 7 µl H2O. Digestion was carried out at 37°C during 15 hours. For X2 program: Reaction was performed in a final volume of 30 µl in which there were 15 µl of PCR product, 3 µl of enzymatic tampon, 1 µl of enzyme, and 11 µl H2O. Digestion was carried out at 37°C during 1 hour. For X3 program, in a final volume of 20 µl, there were 14 µl of PCR product, 5 µl of enzymatic tampon, and 1 µl of enzyme (without H2O). Digestion was carried out at 37°C during 1 hour.

3. Results

3.1 Technical optimization

PCR amplification reactions and conditions were optimized using in Silico PCR tools and gradient melting assays.

One touch-down PCR reaction was finally designed for a simultaneous amplification of the three markers. The optimization of the PCR Touch-Down program and the optimization of the digestion protocol are shown in **Figure 3**.

It was possible via Nebcutter to validate virtual rs4819523 and rs5748411 digestion profiles, using respectively HaeIII and BsrI restriction enzymes. However, the digestion of rs4680 by BclI, as described by Huber (2014) [11], was not possible and another enzyme was then considered. Indeed, the verification of the digestion profile of rs4680 by BclI, showed both manually and by Nebcutter, confirmed the absence of the restriction site at the level of the sequence of the amplicons generated in silico PCR for both the ancestral G allele and the A allele. The design of other enzymes via the same Biolabs Nebcutter tool was then performed. Three other enzymes can be used, FatI, CviAII, and NLaIII, which are isoschizomers (**Figure 4**).

They highlight the selected SNP. The digestion of the rs4680 by CviAII gives the homozygous alleles GG (2 restriction sites and 3 bands) and AA (3 restriction sites and 4 bands). The heterozygous profile is the combination of the two previous ones (**Table 6**).

Fragments generated after restriction/digestion were visualized on 4% agarose gel. **Figure 3** shows the electrophoresis profile of amplification product of rs4819523 using a melt gradient (**Figure 3A**), the electrophoresis profile of the PCR products of the three selected STS (rs4819523, rs4680, and rs5748411) (**Figure 3B**), and finally the electrophoresis profile of the PCR-RFLP products of the rs4819523 using for the digestion, the HaeIII restriction enzyme (**Figure 3C**).

3.1.1 Application

The results of the 22q11.2 exclusion using the optimized PCR-RFLP technique in our sample of 10 patients known as having or not a 22q11 microdeletion are shown in **Table** 7.

Figure 5 shows the PCR-RFLP electrophoresis profile of rs4819523 digested amplicons in our 10 patients as well as the rs4819523 alleles of all patients (**Figure 5** and **Table** 8). **Table 8** Shows the genotypes of the rs4819523 SNP of all patients (**Table 8**).

Application of our technique for our 10 patients involving two deleted and eight undeleted patients proved that the PCR-RFLP 22q11.2 exclusion results were concordant with FISH/MLPA results (**Figure 6**). The status of the deleted subjects was homozygous while the profile of the undeleted ones was heterozygous. PCR-RFLP

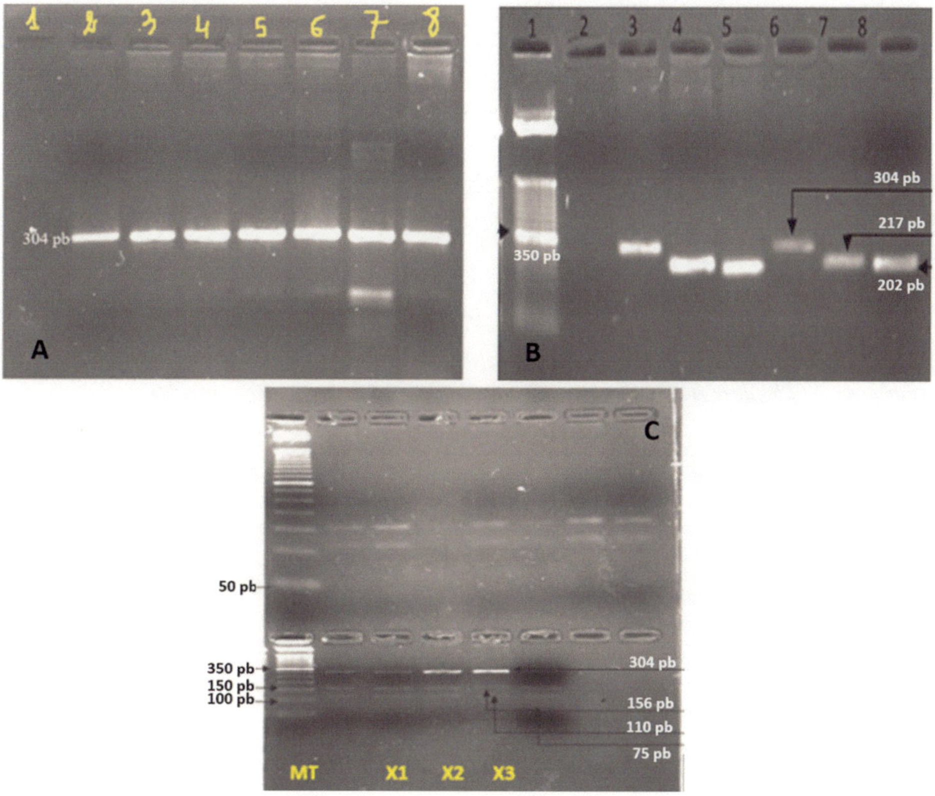

Figure 3.
A: Electrophoresis profile of amplification product of rs4819523 using a melt gradient B: Electrophoresis profile of the PCR products of the 3 selected STS C: Electrophoresis profile of the PCR-RFLP products of the rs4819523 using for the digestion, the HaeIII restriction enzyme. A-legend: Line 1: H2O, line 2: Tm 64,6°C, line 3: Tm 63,7°C, line 4: Tm 62,1°C, line 5: Tm 60,2°C, line 6: Tm 57,5°C, line 7: Tm 55°C, line 8: Amplification using touch-down PCR program B-legend: Line 1: 50 bp ladder DNA marker, line 2: H2O, lines 3,4, and 5: DNA control for the three SNP rs4819523 (304 bp), rs4680 (217 bp), and rs5748411 (202 bp) C-legend: Digestion of amplification product of a DNA control sample for rs4819523 using HaeIII permits to obtain three fragments: 156pb, 110pb, 75pb. This profile is compatible with the heterozygous status C/G. MT: 50 bp ladder DNA marker, X1, X2, X3 are three digestion programs.

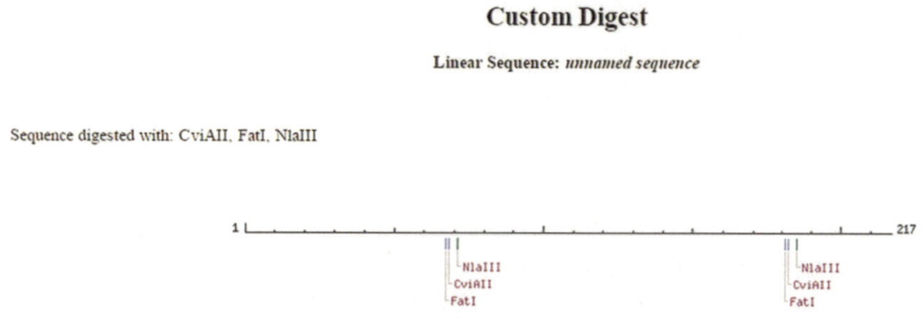

Figure 4.
Restriction enzymes that cut the rs4680.

analysis of rs4819523 showed heterozygote status C/G in the eight undeleted patients with a consequent exclusion of the microdeletion 22q11. Comparatively, in microdeleted patients which was already confirmed by FISH (**Figure 6**) or MLPA,

SNP	Restriction Enzyme	Restriction Sequence	Cleavage Site
rs4680	CviAII	CATG	5′...C↓ATG...3′ 3′...G TA↑C...5′

Cleavage fragments in the presence of the cleavage site	Cleavage fragments in the absence of the cleavage site
Allele A: 4 bands 96pb 69pb 34pb 18pb	Allele G: 3 bands 114pb 69pb 34pb

Table 6.
Electrophoretic profiles after rs4680 digestion using CviAII restriction enzyme.

No	Sex	Age range (years)	Congenital Heart Disease	Karyotype	FISH/MLPA results of 22q11 microdeletion	PCR-RFLP exclusion of microdeletion 22q11
1	F	[2–5]	Double outlet right ventricle	46,XX	Negative	Positive
2	M	[2–5]	Coarctation of Aorta	46,XY	Negative	Positive
3	F	[2–5]	Pulmonic Valve Stenosis	46,XX	Negative	Positive
4	F	[5–10]	Tetralogy of Fallot	46,XX	Positive	Negative
5	M	[0–2]	Coarctation of Aorta/ Persistent Ductus Arteriosus	46,XY	Negative	Positive
6	F	[5–10]	Tetralogy Of Fallot	46,XX	Negative	Positive
7	M	[0–2]	Coarctation of Aorta	Failure of culture	Negative (MLPA)	Positive
8	M	[0–2]	Tetralogy Of Fallot	46,XY	Negative	Positive
9	F	[0–2]	Atrial septal defect/ Persistent Ductus Arteriosus	Failure of culture	Negative (MLPA)	Positive
10	M	[10–15]	Muscular Ventricular Septal Defect	45,XY,-22 der (11)t(11;22) (q23;q11)	Positive	Negative

Table 7.
PCR-RFLP results for exclusion of 22q11 microdeletion in our 10 Tunisian CHD patients.

PCR-RFLP analysis of rs4819523 showed a homozygous G/G status in patient P4 and a homozygous C/C status in the one of P10 (**Table 8** and **Figure 5**).

4. Discussion

Microdeletions and microduplications, known as segmental aneusomies are defined by the presence of a sub-cytogenetic chromosomal anomaly, less than 5 Mb in

Patient number	rs4819523 alleles	Genotype
P1	C/G	Heterozygote
P2	C/G	Heterozygote
P3	C/G	Heterozygote
P4	**G/G**	**Homozygote**
P5	C/G	Heterozygote
P6	C/G	Heterozygote
P7	C/G	Heterozygote
P8	C/G	Heterozygote
P9	C/G	Heterozygote
P10	**C/C**	**Homozygote (ancestral allele)**

Table 8.
rs4819523 genotypes of our 10 Tunisian CHD patients.

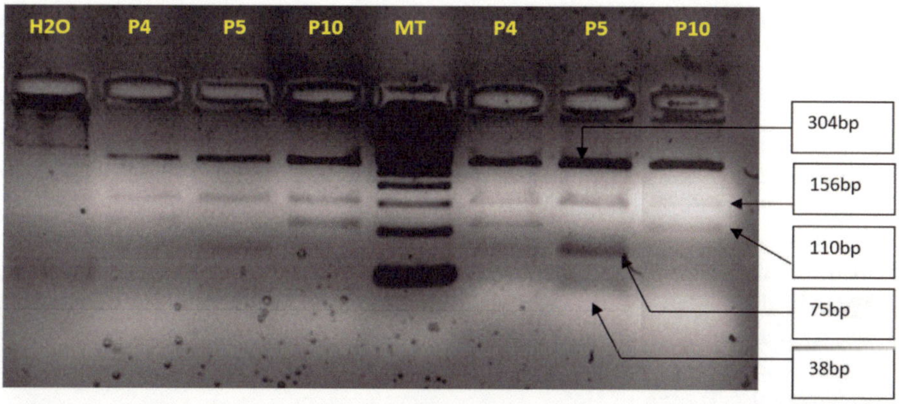

Legend: Profile P3: Heterozygous C/G with four fragments 156bp, 110bp,75bp and 38bp; Profile P4: Homozygous G/G with three fragments 156bp, 75bp and 38bp; Profile P10: Homozygous C/C with three fragments 156bp, 1bp and 38bp.

Figure 5.
RFLP electrophoresis profile of rs4819523 amplicons in the 10 patients. Profile P3: Heterozygous C/G with four fragments 156 bp, 110 bp, 75 bp, and 38 bp; profile P4: Homozygous G/G with three fragments 156 bp, 75 bp, and 38 bp; profile P10: Homozygous C/C with three fragments 156 bp, 1 bp, and 38 bp.

size, classically detectable by the molecular cytogenetic technique of fluorescent in situ hybridization (FISH) [12, 13].

Molecular cytogenetics has made it possible to link microdeletions to syndromes that have long been described clinically but whose genetic origin was suspected. The two oldest entities studied are the 22q11 microdeletion syndrome associated with Digeorge (DG: MIM 188400) and velocardiofacial (VCF: MIM 192430) syndromes and the 7q11 microdeletion syndrome associated with Williams syndrome (WBS: MIM 194050) [14, 15]. It should be noted that there are several synonyms for DG syndrome, namely CATCH 22, Shprintzen syndrome, Takao syndrome, and conotruncal anomaly face (CAF) syndrome [16–18].

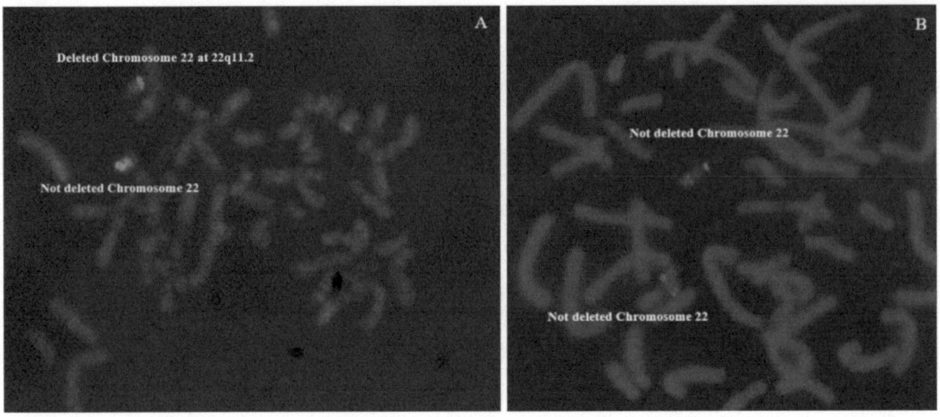

Figure 6.
FISH results showing deleted (A) and not deleted (B) patients (P4 and P6 respectively) (12).

Mechanistically, microdeletions and microduplications occur during meiosis as a result of illegal crossovers with mismatches and non-allelic homologous recombination (NAHR) between DNA sequences with high homology. These are in particular segmental duplications or LCRs for Low Copy Repeats [19]. These are sequences of 10 to 300 kb, with more than 95% homology, often associated with Alu sequences. They are located particularly in the pericentromeric and subtelomeric regions and represent 5 to 10% of the genome, which explains the presence of recombination hotspots. NAHR during meiosis with LCRs in direct orientation leads to deletions/duplications.

Several genes located and cloned in these regions are specifically responsible for certain phenotypes, in particular by haploinsufficiency mechanisms [20]. Recent advances in molecular biology have led to the discovery of a progressively increasing number of microdeletional and microduplicational syndromes through the search for imbalances in the regions delimited by the LCRs using CGH array [21, 22]. The same advances have made it possible to broaden the spectrum of techniques enabling their detection and screening on a large scale, namely CGH array, real-time PCR, and the MLPA technique [20, 23].

The existence of two recurrent types of 22q11 microdeletion has been demonstrated. The first deletion extends over 3 Mb with a commonly deleted region or TDR found in 90% of patients. The second, always proximal, extends over 1.5 Mb and is found in 7% of patients [24, 25]. Both deletions occur by NAHR between LCR22 (A, B, C, and D) with LCRA/D and LCRA/B respectively [26]. Atypical deletions, often distal, sparing the minimal critical region, which is the smallest region responsible for the DG phenotype in its absence, were also reported.

22q11.2 microdeletions, now correspond, thanks to the advent of high-resolution molecular techniques, to a very heterogeneous genomic entity, that of CNVs, linked to recurrent rearrangements in the 22q11. Two region, secondary to too frequent non-allelic homologous recombination due to the richness of the region in LCRs [24]. In fact, mapping of the 22q11 region has revealed the presence of 8 LCR22s identified at the 22q11.2 region, named LCR22-A to LCR22-H or LCR22-2 to LCR22-8 [21, 22]. The four centromeric LCR22s from LCR-A to LCR-D, known for a long time, are involved in the proximal classical recurrent microdeletions specific to DG/VCF syndromes and in reciprocal 22q11.21 microduplication (MIM 608363). The 4 other LCR22 with telomeric disposition, named LCR22-E to LCR22-H located in the region immediately distal to 22q11.21 TDR (i.e. 22q11.22-q11.23), are involved in recurrent distal microdeletions associated with particular phenotypes (MIM 611867) as well as in reciprocal microduplications.

Study of the LCR22-D, −E, and −F region showed the presence of the BLCR module in each of these 3 LCR22s, suggesting the involvement of the NAHR mechanism in distal 22q11.2 microdeletions. In this particular region, Mikhail et al. (2014) demonstrated the presence of microdeletions variable in size, challenging the notion that microdeletion in this region is associated with a single clinical entity [21, 24]. The same author proposed a new classification of distal 22q11.2 microdeletions into three major types I, II, and III emanating from three respective types of NAHR: LCR22-D/−E or -F; LCR22-E/-F and LCR22-F/−G. Molecular studies based on high-resolution techniques have also identified 22q11.2 microdeletions with atypical breakpoints that do not involve any of the known LCR22s [21, 24]. These authors suggested the involvement of SINE/Alu elements and highlighted the extreme complexity of the 22q11.2 genomic region, its rearrangements, and relative phenotypes [27].

Microdeletion and microduplication syndromes are currently referred to as CNVs [21]. A CNV for recurrent copy-number variations, is defined as a segment of DNA whose size varies from 1 kb to several Mb and whose copy number is different from that of a control for which it is fixed at 2. These polymorphisms are distributed throughout the genome and listed on a website: Database of genomic variants. The number of polymorphisms has grown steadily, reaching thousands, spread across all the chromosomes, and representing around 5% of the human genome. They are included in genes, but also in highly conserved regions of the genome and regions that regulate gene expression. Their effect can take the form of a gene dosage effect, corresponding to deletions and duplications in cytogenetic nomenclature, or be linked to the interruption of a gene.

The more than 40 genes in the 22q11.2 region have been extensively studied to delineate the phenotypes associated with 22q11.2 microdeletions and microduplications and their pathophysiological mechanisms, including neural crest migration abnormalities [28]. The TBX1 gene (MIM 602054) belonging to the proximal 22q11.2 region remains the most incriminated gene in the cardiac manifestations of 22q11.2 microdeletions, with apparently clinical heterogeneity secondary to modifying factors such as gene environment [21, 24].

Despite these molecular advances and the overly broad spectrum of phenotypic manifestations associated with 22q11.2 microdeletions classified into several types, the group of proximal 22q11.2 microdeletions with its two subtypes, common 3 Mb microdeletion and central 1.5 Mb microdeletion, remain those most frequently reported to date. Compared to distal microdeletions, proximal 22q11.2 microdeletions are the entities that most include, in their phenotypic spectrum, cardiac malformations, in particular, conotruncal heart disease (MIM 217095) [11, 12, 23, 29, 30]. According to three landmark studies in which the phenotype of 810 patients with microdeletions at 22q11 was determined, conotruncal heart disease was reported in 75% of cases [31–33]. Tetralogy of Fallot is the most common heart disease, with a prevalence of 22–35%. The frequency of CHD also differs according to the involved clinical syndrome. It should be noted that cardiac involvement seems to be rare in distal microdeletions except for deletions classified as type II according to Mikhail et al. (2014) [24] i.e. those resulting from NAHR LCR22-D/E, where the frequency of CHD exceeds 40% and where the CHD seems to correspond to atrial septal defects (ASD) and ventricular septal defects (VSD), patent ductus arteriosus (PCA), and bicuspid aortic valve. With the exception of PCA, these CHD are rarely involved in proximal microdeletions for which the most frequent (but not exclusive) are conotruncal heart defects.

Fluorescence in situ hybridization (FISH), which remains the gold standard test to determine 22q11.2 microdeletion and detects more than 95% cases, is very expensive, time-consuming, and not widely available. It usually requires chromosomal preparations. Settings, such technique for widespread screening of microdeletions, are limited in the medical background with a shortage of resources [4, 34–36].

Given the financial cost of 22q11.21 microdeletion detection techniques and the impossibility of applying them on a large scale, we have tried in this work to optimize a simple and inexpensive molecular technique using PCR-RFLP in order to allow the exclusion of 22q11.21 microdeletion syndrome. This negative screening is helpful to the reduction of the number of patients who are candidates for the confirmation of the 22q11 syndrome by FISH, MLPA, qPCR, or array CGH. The principle of this exclusion is based on the use of three markers located at 22q11.21, each of which has two alleles. The heterozygosity profiles for these CNV confirm the absence of a microdeletion. Only the homozygous profiles would be likely to be deleted. However, it requires confirmation. Experimentally, the PCR-RFLP technique was firstly developed for the rs4819523 marker and subsequently for the two other ones. After optimization, the technique was applied to 10 patients whose microdeletional status was previously known using the two techniques of reference, notably, FISH and MLPA. All patients had CHD, primarily consisting of conotruncal CHD.

The current study allowed us to validate the technique by showing the heterozygous status for the eight undeleted patients and the homozygous status for the two patients with a confirmed 22q11.2 microdeletion. Accordingly, we have demonstrated that this technique is promising for the 22q11.21 region as well as for other chromosomal regions prone to microdeletions such as the distal 22q11.2 and the 7q11 regions.

For an appropriate application in our population, and may be others, it will be necessary to conduct preliminary population studies in order to identify the polymorphic markers whose characteristics and allelic frequencies are accurately estimated. It requires further improvement, notably, the selection of other biallelic markers to cover more widely the 22q11.21 region of 3 Mb as well as the identification of more polymorphic markers of the selected regions in our Tunisian population, to define the allelic forms and ultimately, to estimate their alleles frequencies.

5. Conclusion

In conclusion, the principle of the proximal 22q11.2 microdeletion, applied with the exclusion technique, is apparently interesting and possibly used for several applications. Furthermore, this principle could be used to target patients with distal microdeletions, which are becoming more and more frequent for an increasingly rapid confirmation. In this vein, it could be used for excluding these distal 22q11.22-q11.23 microdeletions. However, it is necessary to choose the appropriate polymorphic markers for each type of distal microdeletion and to develop the technique. Further studies of the population for the determination of the heterozygosity rate of 22q11 region markers are needed before setting up such interesting methods of deletion screening in our population and may be others.

Acknowledgements

The authors are grateful to Doctor Mohamed El Behi and Doctor Rim Louati for their precious contribution.

Conflict of interest

The authors declare no conflict of interest.

Author details

Nouha Bouayed Abdelmoula[1*], Samir Aloulou[1], Sonda Kammoun[1], Aymen Damak[2], Mona Rekik[1], Saloua Ben Amor[1], Oldez Kaabi[1] and Balkiss Abdelmoula[1]

1 Genomics of Signalopathies at the Service of Precision Medicine UR17ES36/LR23ES07, Faculty of Medicine, University of Sfax, Tunisia

2 Department of Thoracic and Cardiovascular Surgery, Faculty of Medicine, Habib Bourguiba Hospital, Sfax, Tunisia

*Address all correspondence to: nouha_abdelmoulabouayed@yahoo.fr

References

[1] Rooney K, Levy MA, Haghshenas S, Kerkhof J, Rogaia D, Tedesco MG, et al. Identification of a DNA methylation episignature in the 22q11.2 deletion syndrome. International Journal of Molecular Sciences. 2021;**22**(16):8611. DOI: 10.3390/ijms22168611

[2] Molck MC, Vieira TP, Simioni M, Sgardioli IC, dos Santos AP, Xavier AC, et al. Distal 22q11.2 microduplication combined with typical 22q11.2 proximal deletion: A case report. American Journal of Medical Genetics. Part A. 2015;**167A** (1):215-220. DOI: 10.1002/ajmg.a.36809 Epub 2014 Oct 30

[3] McDonald-McGinn DM, Hain HS, Emanuel BS, Zackai EH. 22q11.2 deletion syndrome. In: Adam MP, Ardinger HH, Pagon RA, Wallace SE, LJH B, Gripp KW, Mirzaa GM, Amemiya A, editors. GeneReviews® [Internet]. Seattle (WA): University of Washington, Seattle; 1999 Sep 23 [updated 2020 Feb 27], 1993–2022

[4] Morrow BE, McDonald-McGinn DM, Emanuel BS, Vermeesch JR, Scambler PJ. Molecular genetics of 22q11.2 deletion syndrome. American Journal of Medical Genetics. Part A. 2018;**176**(10):2070-2081. DOI: 10.1002/ajmg.a.40504

[5] Ben-Shachar S, Ou Z, Shaw CA, Belmont JW, Patel MS, Hummel M, et al. 22q11.2 distal deletion: A recurrent genomic disorder distinct from DiGeorge syndrome and velocardiofacial syndrome. American Journal of Human Genetics. 2008;**82**(1):214-221. DOI: 10.1016/j.ajhg.2007.09.014

[6] Vervoort L, Dierckxsens N, Pereboom Z, Capozzi O, Rocchi M, Shaikh TH, et al. 22q11.2 low copy repeats expanded in the human lineage. Frontiers in Genetics. 2021;**12**:706641. DOI: 10.3389/fgene.2021.706641

[7] Carli D, Moroni A, Eleonora DG, Zonta A, Montin D, Licciardi F, et al. Atypical microdeletion 22q11.2 in a patient with tetralogy of Fallot. Journal of Genetics. 2021;**100**:5

[8] Motahari Z, Moody SA, Maynard TM, LaMantia AS. In the line-up: Deleted genes associated with DiGeorge/22q11.2 deletion syndrome: Are they all suspects? Journal of Neurodevelopmental Disorders. 2019;**11(1)**:**7**:1-28. DOI: 10.1186/s11689-019-9267-z

[9] Pereira AC, Corrêa RF, Mota GF, Kim CA, Mesquita SF, Krieger JE. High specificity PCR screening for 22q11.2 microdeletion in three different ethnic groups. Brazilian Journal of Medical and Biological Research. 2003;**36**:1359-1365

[10] Gioli-Pereira L, Pereira AC, Mesquita SM, Lopes AA, Krieger JE. PCR screening for 22q11.2 microdeletion: Development of a new cost-effective diagnostic tool. Clinica Chimica Acta. 2006;**369**:78-81

[11] Huber J, Catarino Peres V, Luz de Castro A, Jeronimo dos Santos T, Fontoura D, Beltraõ L, et al. Molecular screening for 22Q11.2 deletion syndrome in patients with congenital heart disease. Pediatric Cardiology. 2014;**35**:56-62

[12] Gargouri B, Abdelmoula N, Trabelsi Sahnoun I, Gargouri B, Rebaï T, Amouri A. Screening for cytogenetic and molecular chro-mosome rearrangements in Tuni-sian children with Conotruncal heart defects. Laboratory Medi-cine. 2009;**40**:357-361

[13] Abdelmoula NB, Portnoi MF, Vialard F, Amouri A, VandenAkker J,

Taillemite JL. Molecular cytogenetic techniques: principles and advances. Médecine/Sciences. 2000;**16**:1405-1441

[14] Lowery MC, Morris CA, Ewart A, Brothman LJ, Zhu XL, Leonard CO, et al. Strong correlation of elastin deletions, detected by FISH, with Williams syndrome: Evaluation of 235 patients. American Journal of Human Genetics. 1995;**57**:49-53

[15] Schluth-Bolard C, Till M, Edery P, Sanlaville D. Syndromes chromosomiques émergents. Pathologie et Biologie. 2008;**56**:380-387

[16] Driscoll DA, Budarf ML, Emanuel BS. A genetic etiology for DiGeorge syndrome: Consistent deletions and microdeletions of 22q11. American Journal of Human Genetics. 1992;**50**:924-933

[17] Scambler PJ, Kelly D, Lindsay E, Williamson R, Goldberg R, Shprintzen R, et al. Velocardio-facial syndrome associated with chromosome 22 deletions encompassing the DiGeorge locus. Lancet. 1992;**339**: 1138-1139

[18] Burn J, Takao A, Wilson D, Cross I, Momma K, Wadey R, et al. Conotruncal anomaly face syndrome is associated with a deletion within chromosome 22q11. Journal of Medical Genetics. 1993; **30**:822-824

[19] Baumer A, Dutly F, Balmer D, Riegel M, Tükel T, Krajewska-Walasek M, et al. High level of unequal meiotic crossovers at the origin of the 22q11. 2 and 7q11.23 deletions. Human Molecular Genetics. 1998;**7**:887-894

[20] Monteiro FP, Vieira TP, Sgardioli IC, Molck MC, Damiano AP, Souza J, et al. Defining new guidelines for screening the 22q11.2 deletion based on a clinical and dysmorphologic evaluation of 194 individuals and review of the literature. European Journal of Pediatrics. 2013;**172**: 927-945

[21] Mikhail FM. Copy number variations and human genetic disease. Current Opinion in Pediatrics. 2014;**26**:646-652

[22] Béri-Dexheimer M, Bonnet C, Chambon P, Brochet K, Grégoire MJ, Jonveaux P. Microarray-based comparative genomic hybridization in the study of constitutional chromosomal abnormalities. Pathologie-Biologie (Paris). 2007;**55**:13-18

[23] Frigerio M, Passeri E, de Filippis T, Rusconi D, Valaperta R, Carminati M, et al. SNPs and real-time quantitative PCR method for constitutional allelic copy number determination, the VPREB1 marker case. BMC Medical Genetics. 2011;**12**:61

[24] Mikhail FM, Burnside RD, Rush B, Ibrahim J, Godshalk R, Rutledge SL, et al. The recurrent distal 22q11.2 microdeletions are often de novo and do not represent a single clinical entity: A proposed categorization system. Genetics in Medicine. 2014;**16**:92-100

[25] Scambler PJ. The 22q11 deletion syndromes. Human Molecular Genetics. 2000;**9**:2421-2426

[26] Vergés L, Molina O, Geán E, Vidal F, Blanco J. Deletions and duplications of the 22q11.2 region in spermatozoa from DiGeorge/velocardiofacial fathers. Molecular Cytogenetics. 2014;**7**:86

[27] Beaujard MP, Chantot S, Dubois M, Keren B, Carpentier W, Mabboux P, et al. Atypical deletion of 22q11.2: Detection using the FISH TBX1 probe and molecular characterization with high-density SNP arrays. European

Journal of Medical Genetics. 2009;**52**: 321-327

[28] Karayiorgou M, Simon TJ, Gogos JA. 22q11.2 microdeletions: Linking DNA structural variation to brain dysfunction and schizophrenia. Nature Reviews. Neuroscience. 2010;**11**:402-416

[29] Abdelmoula NB. Contribution of molecular cytogenetics in the diagnosis of chromosomal abnormalities. Annales de biologie clinique. 2004;**62**:1-9

[30] Wozniak A, Wolnik-Brzozowska D, Wisniewska M, Glazar R, Materna-Kiryluk A, Moszura T, et al. Frequency of 22q11.2 microdeletion in children with congenital heart defects in western Poland. BMC Pediatrics. 2010;**10**:88

[31] McDonald-McGinn DM, Kirschner R, Goldmuntz E, Sullivan K, Eicher P, Gerdes M, et al. The Philadelphia story: The 22q11.2 deletion: Report on 250 patients. Genetic Counseling. 1999;**10**:11-24

[32] McDonald-McGinn DM, LaRossa D, Goldmuntz E, Sullivan K, Eicher P, Gerdes M, et al. The 22q11.2 deletion: Screening, diagnostic workup, and outcome of results; report on 181 patients. Genetic Testing. 1997;**1**:99-108

[33] Ryan AK, Goodship JA, Wilson DI, Philip N, Levy A, Seidel H, et al. Spectrum of clinical features associated with interstitial chromosome 22q11 deletions: A European collaborative study. Journal of Medical Genetics. 1997; **34**(10):798-804

[34] Tomita-Mitchell A, Mahnke DK, Larson JM, Ghanta S, Feng Y, Simpson PM, et al. Multiplexed quantitative real-time PCR to detect 22q11.2 deletion in patients with congenital heart disease. Physiological Genomics. 2010;**42A**(1):52-60

[35] Maran S, Faten SA, Lim SE, Lai KS, Ibrahim WPW, Ankathil R, et al. Screening of 22q11.2DS using multiplex ligation-dependent probe amplification as an alternative diagnostic method. BioMed Research International. 2020; **2020**:6945730

[36] Campos-Garcia FJ, Castillo-Espinola AM, Medina-Escobedo CE, Zenteno JC, Lara-Riegos JC, Rubio-Zapata H, et al. Screening method for 22q11 deletion syndrome involving the use of TaqMan qPCR for TBX1 in patients with Conotruncal congenital heart disease. Cardiogenetics. 2022;**12**(3):253-260

Chapter 4

Pharmacogenomics – A Prospective Journey towards Precision Medicine

Chrisanne Freeman

Abstract

In personalized medicine, genomic data is utilized to focus on individual reactions to drugs. At the point when a gene variation is related to a specific medication reaction in a patient, there is the potential for settling on clinical choices in light of hereditary factors by changing the dose or picking an alternate drug, for instance. Researchers survey gene variations influencing a person's medication reaction the same way they evaluate gene variations related to certain illnesses: by recognizing hereditary loci related to known drug reactions and afterward testing people whose reaction is obscure. Current methodologies incorporate multi-gene investigation or entire genome single nucleotide polymorphism (SNP) profiles, and these methodologies are simply coming into clinical use for drug revelation and improvement.

Keywords: pharmacokinetics, personalized medicine, single nucleotide polymorphism profile, pharmacogenetics, drug safety

1. Introduction

Pharmacogenetics is the science of how genetic factors affect the variation in drug safety and efficacy between individuals [1]. Regarding the human genome project, pharmacogenomics is another applied study of the entire genome, including genomics and proteomics, for recognizing every human gene, inter-individual and intra-individual variations in articulation, and testing its capability progressively [2]. Customized medication is an incredible chance to take a "one size fits all" approach dealing with diagnostics, medication treatment, and counteraction and transform it into an individualized methodology. Of course, we are all alike, but we are also unique, and this allows us to make individual predictions about disease risk, which can help someone choose a prevention plan that is right for them. Genomics is playing a big role in the development of personalized medicine. Genomics gives us a new window into the differences between us in a very specific molecular way. It likewise permits the chance, in certain cases, of picking the perfect medication at the ideal dosage for the perfect individual rather than the "one size fits all" way to deal with drug treatment. In the end, it will be difficult to see how this will not affect any kind of medicine as we learn more about each person and as many of us find that our entire genomes are being sequenced and available as a ready reckoner to enable that kind of personalized approach. There is much work to be done, however, and it might be the greatest upheaval in medication in seemingly forever. Pharmacogenomics aims to design new drugs and select the best treatment for each individual patient. As a general rule, most pharmacogenetic studies target single genes and their relationships with individual contrasts in drug interactions; at the same time, pharmacogenomics is a science that also deals with genomic interactions between genes in the general variety of drug metabolism and reactions.

2. Pharmacogenomics' - origin and development

Pharmacogenetics existed more than 2000 years before Pythagoras' observations, but enzyme polymorphisms like N-acetyltransferase and G6PD were not discovered until the 1950s [3]. In 1959, Friederich Vogel came up with the term "pharmacogenetics" to describe a new field of study that uses genetics and pharmacological knowledge and techniques to investigate the impact of inherited factors on drug response variability [4]. The subject of pharmacogenetics became too esoteric, and as a result, it declined. Proof for a genetic basis for clinical disorders related to the administration of medications arose in the mid-'50s when antimalarial drugs, for instance, primaquine, were proven to stimulate hemolytic anemia in patients who had a deficiency of glucose-6-phosphate dehydrogenase. Numerous observations of pharmacogenetic-based variations in pharmacokinetics were made in the 1970s following the discovery of the CYP2D6 polymorphism and its impact on drug toxicity and response [5]. This examination brought about a few investigations in view of the utilization of molecular innovations connected to traditional pharmacological phenotypization and hereditary studies in populations that allowed the identification of a few polymorphisms in genes engaged in drug metabolism. The term "pharmacogenomics" was first used in the medical literature towards the end of the 1990s. The European Agency for the Evaluation of Medicinal Products (EMEA) characterizes "pharmacogenetics" as "the investigation of inter-individual variations in DNA sequences connected with drug reactions" and "pharmacogenomics" as "the investigation of the expression of individual genes responsible for disease susceptibility as well as drug toxicity at the cell, tissue, individual, or populace level." The European Agency for the Evaluation of Medicinal Products (EMEA), 2002. The most common meaning of the term, which considers pharmacogenomics to be the evolution of pharmacogenetics on a genomic scale, is in line with this definition. Pharmacogenetics, as a matter of fact, uses genetic innovations to investigate a set number of genes to describe the molecular mechanism of an individual's reaction to drugs, while pharmacogenomics includes the investigation of the entire genome as it connects with drug reactions using high-throughput advancements.

At the drug metabolizing enzyme, transporter, or receptor level, polymorphisms associated with variable drug response have been identified for an increasing number of genes, mostly through a candidate gene approach. The use of genome-wide analysis is leading to the discovery of previously unknown new genes linked to disease and drug response. Albeit a few old and most new medications going onto the market have a "pharmacogenomic track" the clinical significance of pharmacogenomics has been by and large inadequate. For toxicity (such as azathioprine) and efficacy (such as warfarin) purposes, narrow therapeutic index drugs have been the primary focus of clinical translation of pharmacogenetics to date. Pharmacogenetics and genomics will progress through lower-cost, fast entire genome sequencing techniques joined with complex calculations permitting individualized measurement suggestions, but not really their reception. However, the influence of environmental and genetic factors on gene expression changes complicates this. As a result, the translation of pharmacogenetics into "personalized medicine" will be contingent on a variety of elements, such as clinical relevance, interactions between genes and the environment, costs, and education [3].

3. Single nucleotide polymorphisms: its impact on drug metabolism

Genetic variation in drug metabolism is a major factor in the variation in drug toxicity between individuals. Drug-metabolizing enzymes involving single nucleotide polymorphisms create distinct population subgroups with distinct capabilities for drug-transforming reactions. For the majority of enzymes that break down drugs, genetic polymorphisms have been described. Mutations in these enzyme genes cause polymorphisms by reducing, increasing, or eliminating enzyme expression or activity through a variety of molecular mechanisms. In addition, the population contains recessive alleles with a relatively high frequency [6]. Genes assist with building proteins and their enzymes. Many things happen to enzymes, like breaking down (metabolizing) drugs. People who do not respond to medications as well as expected might have genetic differences that make it harder for enzymes to break down a medication or make them stop working. A person's response to a medication may be affected by these genetic variations. A typical dose of a drug can cause side effects or have little to no effect on treating the condition at hand if it is broken down too quickly or too slowly. The drug itself also affects how a person responds to it. For instance, whereas increased breakdown renders drug Y ineffective, increased breakdown causes side effects with drug X. The nucleotide sequence of the majority of genes is thought to have evolved casually over time. Mutations situated in a systematizing locale might prompt the replacement of an amino acid in a particular place in a protein and subsequently influence the function of proteins. Mutations in a regulatory region may alter the expression levels of mRNA and proteins and thus transcriptional and translational mechanisms [7].

A polymorphism is a variation in the DNA sequence that occurs in a population with an allelic frequency of at least 1%, while a mutation is a variation that occurs less frequently. Mutations and polymorphisms arrange for enzymes portrayed by various metabolic movements or receptors with affinity for the drug. They alter the pharmacological response of an individual or, in the case of variations that are particularly prevalent in particular ethnic groups, even of a population [7]. Single-nucleotide polymorphisms (SNPs) are the simplest genetic variants. Genetic mutations may include a few nucleotides or long DNA characteristics. In this instance, they are defined as amplifications, translocations, substitutions, insertions, deletions, and large mutations [7]. SNPs can be roughly divided into four main categories, which are as follows: (i) in the gene's protein coding sequence; (ii) in the gene's regulatory regions (like the promoter, 5′-untranslated region, 3′-untranslated region, and intronic sequences); (iii) at the gene's exon-intron boundaries; and (iv) in the gene's intergenic regions, which are interfering genomic segments that separate genes [8].

Alterations in the structure and function of the encoded proteins, as well as changes in the level of gene expression, are all possible outcomes of these sequence variants. However, there may be no discernible effects on protein function.

Clinical Condition	Genes Associated	Clinical Usage	Reference
Atrial fibrillation	CYP2C9, VKORC1	Dose of Warfarin	Redekop and Mladsi [15]
Breast cancer	HER2	Use of Trastuzumab recommended	Redekop and Mladsi [15]
Epilepsy	HLA-B1502	Use of carbamazepine	Redekop and Mladsi [15]
Chronic myeloid leukemia (CML)	BCR and ABL	Imatinib is recommended	Druker et al. [16]
Cystic fibrosis	G551D, G551D	Ivacaftor is recommended	Ramsey et al. [17]

Table 1.
Examples of genetic variants that influence drug metabolism in clinical conditions.

The inheritance of these alleles by patients receiving standard doses of medication can result in an adverse drug reaction or failure to respond in the latter two scenarios. Such SNPs are potential candidates for drug response-modifying alleles. SNPs in genes' regulatory regions have the potential to influence gene expression regulation [9]. Short sequences (typically 6–20 bases) known as transcriptional regulatory domains are found mostly in the promoter or intronic region of genes and serves as transcription factor (TF) binding sites. SNPs that change the binding site might possibly increase or reduce the binding efficiency of transcription factors, which results in spatial modifications in gene expression or potentially changes in the degree of gene expression. Alternately, SNPs in the promoter region may result in a gain-of-function by introducing novel TF binding properties. For instance, the minor "A" allele of a SNP found in the promoter of the tumor necrosis factor gene makes a new binding site for the OCT-1 TF, prompting increased transcriptional action [10]. In contrast, OCT-1 does not bind to the same promoter that carries the predominant "G" allele [10]. At long last, one more gene in the regulatory region that can be impacted by SNPs is the 5′- or 3′-untranslated area [9].

Post-transcriptional regulation of the mRNA involves either translational repression or changes in mRNA stability in these regions, which are on either end of the transcribed mRNA molecule. The binding of regulatory factors—short non-coding RNA molecules with a length of 19–21 nucleotides—to sequence motifs in the untranslated region of the mRNA acts as a conduit for post-transcriptional control [11, 12]. Alterations in regulatory protein [13] or microRNA binding characteristics [14] have been linked to changes in mRNA stability caused by SNPs targeting these motifs in the 3′-untranslated region. In pharmacogenetics, prototypes are used to describe monogenic traits. They are made up of polymorphisms in a single gene that code for a protein in a drug's effects or metabolism, resulting in varying individual responses (**Table 1**). To be viable, drugs should interact with explicit targets restricted to the plasma, cell layer, or cytoplasm. These effectors can be modified qualitatively in the amino acid sequence or quantitatively (in the levels of gene expression) to cause biological variability as well as genetically determined diseases. The administration of a drug that is safe and effective in the general population may, in either case, have severe side effects in people with the disease gene and manifest a subclinical change in a relatively uncommon but clinically significant syndrome like the long QT syndrome.

4. Drug development utilizing pharmacogenomics

The first step in the process of drug discovery is to identify a potential target that the drug could target. A protein involved in signal transduction, a receptor, a transporter, an enzyme in an important pathway, or any protein produced by a disease can all serve as the target. The number of drug targets after sequencing the human genome was estimated to be around 8000, of which 4990 could actually be acted upon—2329 by antibodies and 794 by drug proteins [18]. 399 molecular targets from 130 protein families have been found through ligand binding studies [19, 20]. These targets are known to vary due to genetic polymorphisms. The effects of drugs that are based on targets with wide polymorphisms can vary. As previously mentioned, polymorphisms in the 2 adrenoceptor gene, for instance, have resulted in responder and non-responder phenotypes [21]. This can prompt conflicting outcomes in the preclinical and clinical examinations that would follow in the event that such a compound is sought after as a drug. Such targets can be eliminated as drug compounds, and other appropriate targets can be chosen. So, targets can be characterized early on using pharmacogenetic and proteomic studies, and suitable drug compounds can be chosen for future investment. Variation in a disease's drug response is typically the result of multiple genes rather than a single gene mutation. The aftereffects of pharmacogenetic studies do not have any significant bearing when utilized clinically, as possibly single gene mutations are considered when, as a matter of fact, multiple genes are involved. In such cases, more than a pharmacogenetic study, it would be proper to do pharmacogenomic investigations looking at single nucleotide polymorphism (SNP) expression and heat maps among patients and controls (**Figure 1**). This can distinguish the hereditary elements related to the disease condition and hence give more current focuses to describe and assess, with the end goal of drug development [22].

At the point when targets are tried and tested, in view of the pharmacogenetic profile of patients and their classification, it gives the feeling that those with poor use limits are restricted in their use. When viewed from a broader perspective, it should be noted that this method only reveals what was missed in the pregenomic era, when clinical trials and clinical practice were poorly explained. By using pharmacogenetic devices and understanding the reasons for unfavorable impacts, the targets that

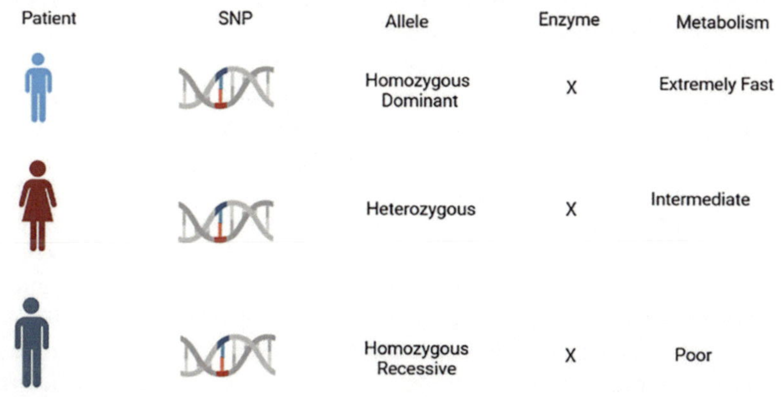

Figure 1.
Consequences of polymorphisms on drug metabolism.

induce morbidity in poor metabolizers can really be forestalled when pharmacogenetics is recommended in clinical practice with suitably directed dosages. Additionally, it must be recognized that the population with poor metabolizing capacity as a result of genetic polymorphisms is only a small, extremely rare subset. The pharmaceutical company avoids developing such a drug if an enzyme polymorphism is found in a larger population. Another worry would be the expense that the patient would incur for pharmacogenetic testing prior to beginning treatment. The expense of genotyping for single nucleotide polymorphisms may not be reasonable in many developing and immature nations. However, as technology advances, this price may decrease in the near future. The cost of genotyping 1000 DNA samples would be 0.3 USD per genotype, as previously mentioned. Yet, when the expense is determined for a solitary patient example, it adds up to in excess of 130 USD, which includes the cost of the probe as well [23]. Consequently, it appears that genotyping is financially effective provided that it is utilized for a larger scope, which would be the case in the event that it is significant for therapeutic purposes.

5. Clinical interpretations

Genetic biomarkers have been the subject of numerous studies on their clinical utility in drug therapy [24, 25] and their potential to partially replace therapeutic drug level monitoring [26, 27]. Additionally, combining genetic biomarkers and drug levels could further guide optimal dosing, such as for warfarin. It is intuitive to assume that drug response and variants in genes that encode drug-metabolizing enzymes, membrane transporters, and receptors are causally linked. Multiple components of the signaling pathway for drug receptors have the potential to introduce variation in drug response, reducing the impact of drug receptor variants alone. The essential objectives of pharmacogenomic biomarkers are the choice of a reasonable treatment methodology or an endless drug measurement routine. Poor metabolizers, or "null alleles," should avoid drugs that are mostly metabolized by a single enzyme; however, partially increased or decreased activity can be used to adjust dosages. Even when taking into account the most recent discoveries in genomics, biomarker predictions of graded enzyme activity frequently exhibit large variations, reducing their clinical utility, such as for CYP2D6 variants. Again, a lot of personal factors need to be taken into account when choosing a treatment.

The implementation of pharmacogenetic and genomic biomarkers into clinical practice is difficult and constrained by numerous factors. Most importantly, there is a lot of variation among patients, so any biomarker can only predict a small portion of disease risk or treatment aftereffects (**Figure 2**). Hence, the overall effect of inter-patient variance, along with cost, are the fundamental models that characterize cost–benefit proportions and decide clinical plausibility. Drugs that are linked to genetic biomarkers are listed in the FDA Pharmacogenomic Biomarkers in Drug Labels, which includes links to complete public drug label information, such as the type of genetic variant and allele frequencies between ethnic groups [28]. As an element of a biomarker's clinical effect, drug-biomarker matches are either given on specialists advice only or further featured with a boxed advance notice when adverse reactions can be extremely severe. For instance, the main active metabolite endoxifen is activated by CYP2D6 when tamoxifen is used to treat breast cancer. Consequently, poor CYP2D6 metabolizers have a lower chance of responding and may require additional treatments. However, the involvement of a number of

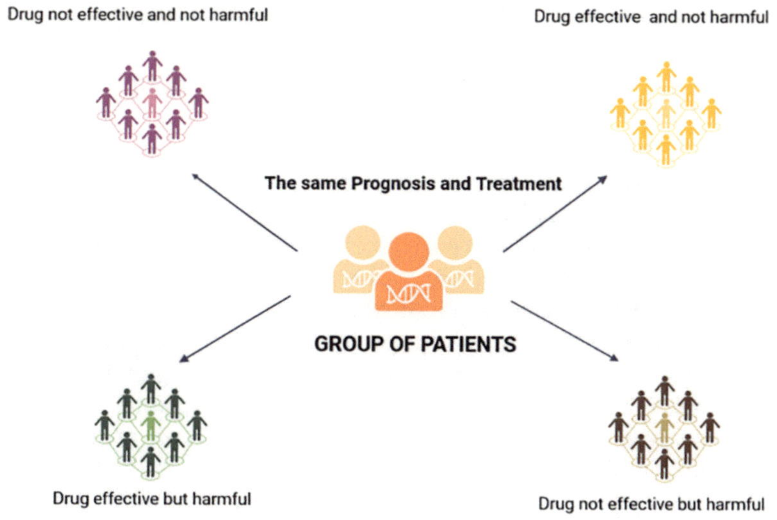

Figure 2.
Drug response variability between individuals.

metabolizing enzymes, dietary factors, and patients' compliance all contribute to variations in response. Further examination is expected to improve the clinical utility of CYP2D6-directed treatment [29].

6. Pharmacogenomics' applications - a network for complex diseases

For personalized treatment plans, precision medicine now incorporates lifestyle, age, behaviors, and polypharmacy regimens in addition to genomic data. A significant part of customized medication is recognizing genes that can impact the metabolism of a drug, which are referred to as "pharmacogenes". Proven and factual variations of specific genes, like those coding for the cytochrome P450 catalyst superfamily, have been shown to influence drug vulnerability and, as a rule, increase side effects. Some people have deletions or substitutions that cause some enzymes to have low, high, or no activity at all, resulting in variable drug metabolism. This complicates the genetic influence. Poor reimbursement from third-party funders, a lack of clinician familiarity with personalized treatment, an inadequate workflow agenda, and a lack of organization in reporting are all obstacles to personalized medicine, despite its growing use in therapeutic areas like psychiatry, cardiology, and pain management. Healthcare administrators require electronic medical recording systems, proper orientation, and a constant genotype and phenotype reporting system, while funders and providers have legitimately requested sufficient published data to support the clinical utility of pharmacogenomics. About half of Americans say they take at least one prescription medication, according to data from the National Centre for Health Statistics [30]. In addition, 28% of people receive prescriptions for three or more medications, and approximately 13% report taking five or more medications in the past 45 days [31]. Altogether, roughly 4.8 billion remedies were filled at drug stores in the US in 2019 [32]. Mostly, numerous patients will encounter adverse consequences of prescription use, either as unfavorable adverse reactions to medication or drug inefficacy, requiring supervision in treatment management [33]. It has been more difficult to demonstrate the clinical significance of prospective, randomized clinical trials despite the abundance of scientific evidence for individual gene-drug associations. Personalized guided treatment and conventional treatment systems have typically been compared in these individual trials [34–36].

These trials are dependent on certain major factors: For instance, a random report with randomized subjects cannot be ethically generated, and these trials cannot be truly triple-blinded. Additionally, patients randomized to conventional treatment receive an effective trial that is proven to be safe and approved by the U.S. FDA. In cases like depression, where superiority over a placebo must be demonstrated, these trials differ significantly from typical industry-sponsored drug trials. At last, the clinician-agents in personalized medicine preliminaries are not typically committed to following the treatment routine characterized by pharmacogenomics. As a result, it should not come as a surprise that such randomized conventional trials lack power or are noted for only marginal outcomes of improvement. There are two reasons why pharmacogenomics will probably be used more frequently in the near future. First, the data will force funders, medical associations, and regulatory agencies to take this approach. Second, and as significant, new generations of doctors and partnered medical care suppliers will turn out to be more educated about the individual, monetary, wellbeing, and cultural ramifications of personalized medicine. Besides, as doctors take up pharmocogenomics in their practices, drug specialists will thusly have to assume a focal role in drug usage and clinical translation [37]. More current and more affordable genetic innovations will further develop the money-saving advantages of personalized medicine. At last, more funders will perceive the financial advantages of pharmacocogenomics and will repay for its utilization.

7. Pharmacogenomics - execution, regulations and ethics

Endless pharmacogenomics biomarker tests keep arising, yet endorsement by administrative organizations and acknowledgment by medical coverage organizations and foundations require legitimate proof of clinical significance. The FDA has established requirements that must be met before a product can be approved. Often, clinical trials are used to compare efficacy in the targeted population to a comparable biomarker-negative control. Cost–benefit analyses are necessary to demonstrate that the biomarker-drug combination is superior or that the drug in and of itself would not meet FDA approval criteria for use in the general patient population, as was the case with trastuzumab in the treatment of breast cancer [38]. Pharmacoeconomics evaluates the money-saving advantage proportions for therapeutic purposes, including the utilization of pharmacogenomics compared with medical care, not surprisingly. The utilization of next-generation sequencing to direct remedial choices raises unexpected issues, for example, coincidental discoveries of pathogenic variations [39]. In the end, it's important to think about other approaches that might be less expensive than pharmacogenomics when making a clinical decision. When high-dose simvastatin is required to lower cholesterol levels, causing potential toxicity, prescribing alternative statins eliminates the requirement for SLCO1B1 genetic data, for instance, in the case of simvastatin and genetic variants of SLCO1B1. Genuine information is as yet meager and will require a normalized, certifiable proof plan. A few studies have addressed evidence review standards, payer participation in study design, and provider and payer education regarding NGS [39]. Utilization of hereditary information summons delicate issues in regards to privacy, abuse by outsiders, and inquiries with pertinence to the patient's family, assuming that malicious changes are found. In the United States, the Genetic Information Nondiscrimination Act (GINA) was passed in 2008, ensuring that genetic information would not be used in decisions regarding health insurance or employment. This alleviated concerns regarding the misuse of genetic information and discrimination. Genetic information will now have a foothold in health care thanks to this important civil rights bill. For patients, the option to be aware of or protected from discoveries of pernicious variations should be explained by marked informed consent in clinical investigations.

The real moral standard in medication is to cause no damage and to stick to the vital core values of equality, justice, and beneficence. In the United States, there are significant disparities in access to health care among various populations. The COVID pandemic is only partially to blame for the decline in life expectancy that has been occurring for a number of years, despite advances in medical science. Black people and indigenous Americans have been disproportionately affected. Minority populations' social exclusion from health care remains a serious issue. Bracic et al. [40], contend that efforts to eliminate such disparities in the context of social behaviors performed by members of the dominant group and members of the minoritized group frequently serve to perpetuate "exclusion cycles." Extending the utilization of enormous information and man-made intelligence-based frameworks in medication to tackle these issues conveys a risk of building up such cycles when established on one-sided data [40]. However, through the use of widely accessible personal electronic health care records, personalized medicine inevitably moves towards convergence and integration of all medical, genomic, personal, cultural, and socioeconomic factors [41]. It is unclear to what extent this method can replace the intuitive judgments of seasoned healthcare professionals, but it is essential to observe beneficence, justice, and equality. Life expectancy has decreased in the United States over the past few years as a result of the two leading causes of death, heart disease and cancer, despite advancements in medicine. Genomics medication guarantees further developed results, yet it must be reasonably coordinated into normal clinical practice to turn out to be completely compelling.

8. Conclusion

Through enhanced use of existing biomarkers and the detection of early genomic and epigenomic events in disease development, particularly carcinogenesis, knowledge of personalized medicine enables earlier disease detection. Preventative medicine is the primary focus of this approach, which encourages proactive rather than reactive actions. This approach delays or forestalls the need to apply more extreme medicines, which are generally less endured and have expanded personal satisfaction and monetary contemplations. Globally, rising healthcare costs, particularly for end-of-life care, have increased pressure on government-funded healthcare systems. Precision medicine might make existing treatments work better and get rid of the problems that come with other methods. Precision medicine is a growing field of medical services where a doctor can choose a therapy in view of a patient's hereditary or genetic profile that may not just limit harmful incidental effects and assure more success, but can also be less practical and an 'experimentation' way to deal with sickness therapy. The "trial-and-error" non-precision medicine approach, which is less effective and can result in drug toxicity, severe side effects, reactive treatment, and misdiagnosis, continues to drive up healthcare costs. A more unified treatment strategy tailored to each individual and their genome will emerge as a result of advances in personalized medicine. Customized medication might furnish better conclusions with prior intercession, more productive medication advancement, and more designated treatments.

To conclude, progress in personalized health care necessitates the convergence of a number of different fields and technologies in order to uncover connections between various components that are intertwined and influence one another. Seeing such intricacies might arise out of man-made brainpower and artificial intelligence later on. Basic genomics research must strive to integrate key aspects of personalized medicine in order to improve clinical translation, as pharmacogenomics serves as an essential link.

Conflict of interest

The author declares no conflict of interest.

Author details

Chrisanne Freeman[1,2]

1 Department of Zoology and Research Centre, Lady Doak College, Madurai, Tamil Nadu, India

2 Department of Biotechnology, Bishop Heber College, Tiruchirappalli, Tamil Nadu, India

*Address all correspondence to: chrisanne@ldc.edu.in

References

[1] Meyer UA. Pharmacogenetics - five decades of therapeutic lessons from genetic diversity. Nature Reviews. Genetics. 2004;**5**(9):669-676. DOI: 10.1038/nrg1428

[2] Evans WE, Relling MV. Moving towards individualized medicine with pharmacogenomics. Nature. 2004;**429**(6990):464-468. DOI: 10.1038/nature02626

[3] Somogy A. Evolution of pharmacogenomics. Proceedings of the Western Pharmacology Society. 2008;**51**:1-4

[4] Mini E, Nobili S. Pharmacogenetics: Implementing personalized medicine. Clinical Cases in Mineral and Bone Metabolism. 2009;**6**(1):17-24

[5] Carson PE, Flanagan CL, Ickes CE, et al. Enzymatic deficiency in primaquine-sensitive erythrocytes. Science. 1956;**124**:484-485

[6] Meyer UA, Zanger UM. Molecular mechanisms of genetic polymorphisms of drug metabolism. Annual Review of Pharmacology and Toxicology. 1997;**37**:269-296. DOI: 10.1146/annurev.pharmtox.37.1.269

[7] Relling MV, Giacomini KM. Pharmacogenetics, chapter 4. In: Brunton LL, Lazo GS, Parker KL, editors. Goodman & Gilman's the Pharmacological Basis of Therapeutics. New York: XI edizione, McGraw-Hill Medical Publishing Division; 2008. pp. 93-115

[8] Lee NH. Pharmacogenetics of drug metabolizing enzymes and transporters: Effects on pharmacokinetics and pharmacodynamics of anticancer agents. Anti-Cancer Agents in Medicinal Chemistry. 2010;**10**(8):583-592. DOI: 10.2174/187152010794474019

[9] Prokunina L, Alarcon-Riquelme ME. Regulatory SNPs in complex diseases: Their identification and functional validation. Expert Reviews in Molecular Medicine. 2004;**6**:1-15

[10] Knight JC, Udalova I, Hill AV, Greenwood BM, Peshu N, Marsh K, et al. A polymorphism that affects OCT-1 binding to the TNF promoter region is associated with severe malaria. Nature Genetics. 1999;**22**:145-150

[11] Filipowicz W, Bhattacharyya SN, Sonenberg N. Mechanisms of post-transcriptional regulation by microRNAs: Are the answers in sight? Nature Reviews. 2008;**9**:102-114

[12] Keene JD. RNA regulons: Coordination of post-transcriptional events. Nature Reviews. 2007;**8**:533-543

[13] Fritz DT, Jiang S, Xu J, Rogers MB. A polymorphism in a conserved posttranscriptional regulatory motif alters bone morphogenetic protein 2 (BMP2) RNA: Protein interactions. Molecular Endocrinology (Baltimore, Md). 2006;**20**:1574-1586

[14] Chin LJ, Ratner E, Leng S, Zhai R, Nallur S, Babar I, et al. A SNP in a let-7 microRNA complementary site in the KRAS 3′ untranslated region increases non-small cell lung cancer risk. Cancer Research. 2008;**68**:8535-8540

[15] Redekop WK, Mladsi D. The faces of personalized medicine: A framework for understanding its meaning and scope. Value in Health. 2013;**16** (Suppl. 6):S4-S9

[16] Druker BJ, Guilhot F, O'Brien SG, et al. Five-year follow-up of patients receiving imatinib for chronic myeloid leukemia. The New England Journal of Medicine. 2006;**355**(23):2408-2417

[17] Ramsey BW, Davies J, NG ME, et al. A CFTR potentiator in patients with cystic fibrosis and the G551D mutation. The New England Journal of Medicine. 2011;**365**:1663-1672

[18] Burgess J, Golden J. Cracking the druggable genome. Bio-IT World [online]. 2002. Available from: http://www.bio-itworld.com/archive/100902/firstbase.html [Accessed: March 7, 2008].

[19] Hopkins AL, Groom CR. The druggable genome. Nature Reviews. Drug Discovery. 2002;**1**:727-730

[20] Imming P, Sinning C, Meyer A. Drugs, their targets and the nature and number of drug targets. Nature Reviews. Drug Discovery. 2006;**5**:821-834

[21] Durham LK, Webb SM, Milos PM, Clary CM, Seymour AB. The serotonin transporter polymorphism, 5HTTLPR, is associated with a faster response time to sertraline in an elderly population with major depressive disorder. Psychopharmacology. 2004;**174**:525-529

[22] McCarthy AD, Kennedy JL, Middleton LT. Pharmacogenetics in drug development. Philosophical Transactions of the Royal Society of London. Series B, Biological Sciences. 2005;**360**(15):79-88

[23] Preventiongenetics. Research genotyping [webpage in internet] 2008. Available from: http://www.preventiongenetics.com/resgeno/listofprices.htm [Accessed: March 6, 2008].

[24] Garcia-Gonzalez X, Cabaleiro T, Herrero MJ, McLeod H, Lopez-Fernandez LA. Clinical implementation of pharmacogenetics. Drug Metabolism and Personalized Therapy. 2016;**31**:9-16

[25] Relling MV, Evans WE. Pharmacogenomics in the clinic. Nature. 2015;**526**:343-350

[26] Haidar CE, Crews KR, Hoffman JM, Relling MV, Caudle KE. Advancing pharmacogenomics from single-gene to preemptive testing. Annual Review of Genomics and Human Genetics. 2022;**23**:449-473

[27] Salman B, Al-Khabori M. Applications and challenges in therapeutic drug monitoring of cancer treatment: A review. Journal of Oncology Pharmacy Practice. 2021;**27**:693-701

[28] Mehta D, Uber R, Ingle T, Li C, Liu Z, Thakkar S, et al. Study of pharmacogenomic information in FDA-approved drug labeling to facilitate application of precision medicine. Drug Discovery Today. 2020;**25**:813-820

[29] Chan CWH, Law BMH, So WKW, Chow KM, Waye MMY. Pharmacogenomics of breast cancer: Highlighting CYP2D6 and tamoxifen. Journal of Cancer Research and Clinical Oncology. 2020;**146**:1395-1404

[30] CDC. Therapeutic Drug Use. CDC, National Center for Health Statistics. 2022. Available from: https://www.cdc.gov/nchs/fastats/drug-use-therapeutic.htm

[31] Dowd D, Williams G, Van Dorn D, et al. Predicting drug-drug and drug-gene interactions in a community pharmacy population. The American Journal of Managed Care. 2022;**28**(11):566-571. DOI: 10.37765/ajmc.2022.89259

[32] Kaiser Family Foundation. Number of Retail Prescription Drugs Filled at

Pharmacies by Payer. Washington: Kaiser Family Foundation; 2020

[33] Schärfe CPI, Tremmel R, Schwab M, Kohlbacher O, Marks DS. Genetic variation in human drug-related genes. Genome Medicine. 2017;**9**:117

[34] Oslin DW et al. Effect of Pharmacogenomic testing for drug-gene interactions on medication selection and remission of symptoms in major depressive disorder: The PRIME care randomized clinical trial. JAMA. 2022;**328**:151-161

[35] Perlis RH, Dowd D, Fava M, Lencz T, Krause DS. Randomized, controlled, participant- and rater-blind trial of pharmacogenomic test-guided treatment versus treatment as usual for major depressive disorder. Depression and Anxiety. 2020;**37**:834-841

[36] Greden JF et al. Impact of pharmacogenomics on clinical outcomes in major depressive disorder in the GUIDED trial: A large, patient- and rater-blinded, randomized, controlled study. Journal of Psychiatric Research. 2019;**111**:59-67

[37] Haidar C-E, Petry N, Oxencis C, Douglas JS, Hoffman JM. ASHP statement on the Pharmacist's role in clinical pharmacogenomics. American Journal of Health-System Pharmacy. 2022;**79**:704-707

[38] Jacobs AT, Martinez Castaneda-Cruz D, Rose MM, Connelly L. Targeted therapy for breast cancer: An overview of drug classes and outcomes. Biochemical Pharmacology. 2022;**204**:115209

[39] Deverka PA, Douglas MP, Phillips KA. Use of real-world evidence in US payer coverage decision-making for next-generation sequencing-based tests: Challenges, Opportunities and Potential Solutions. Value Health. 2020;**23**:540-550

[40] Bracic A, Callier SL, Price WN. Exclusion cycles: Reinforcing disparities in medicine. Science. 2022;**377**:1158-1160

[41] Hicks JK, Dunnenberger HM, Gumpper KF, Haidar CE, Hoffman JM. Integrating pharmacogenomics into electronic health records with clinical decision support. American Journal of Health-System Pharmacy. 2016;**73**:1967-1976

Section 3

SNP Analysis Challenges in the Post-GWAS Era

Chapter 5

Genome-Wide Association in the Mitochondrial Genome Identifies Two Novel Genes Involved in Diabetes Mellitus Type 2

Julio Alejandro Valdez, Pedro Mayorga, Rafael Villa Angulo and Carlos Villa Angulo

Abstract

Diabetes Mellitus Type 2 (DM2) is a complex and multifaceted disorder currently listed as one of the epidemics of the twenty-first century due to its prevalence and the adverse cardiovascular effects it causes. This chapter examines the relationships between base-pair positions in human mitochondrial genome and type 2 diabetes. The data included 510 complete mitochondrial genomes, of which 437 belonged to individuals with type 2 diabetes and 73 to healthy individuals. An alignment algorithm allowed inspecting and choosing a region with optional positions for analysis, a principal component analysis permitted viewing the data structure, and after a regression analysis, we declared three base-pair positions associated to DM2. Upon examination of the genome annotation, three genes were identified as potential candidates for association, one of which was previously linked to type 2 diabetes according to previous studies. This chapter offers further proof of a possible genetic link between type 2 diabetes and metabolic syndrome.

Keywords: genome-wide association study (GWAS), type 2 diabetes mellitus (DM2), logistic regression, principal component analysis (PCA), risk factors

1. Introduction

A group of metabolic disorders known as diabetes mellitus are characterized by chronic hyperglycemia and can be caused by problems with insulin secretion, insulin action, or both. Alterations to the lipid and protein metabolism coexist with hyperglycemia. Long-term sustained hyperglycemia is linked to damage, dysfunction, and failure of many different organs and systems, particularly the heart, blood vessels, nerves, and the retina [1].

There are several types of diabetes and other categories of glucose intolerance. **Type 1 diabetes mellitus (DM1):** Its hallmark is autoimmune destruction of the β cell, which causes absolute insulin deficiency, and a tendency to ketoacidosis. Such destruction in a high percentage is mediated by the immune system, which can be evidenced by the determination of antibodies: Anti-GAD (anti glutamate decarboxylase), anti-insulin, and against the islet cell, with a strong association with the specific DQ-A alleles and DQ-B of the major histocompatibility complex (HLA). DM1 can also be of idiopathic origin, where the measurement of the aforementioned antibodies gives negative results [1]. It usually manifests itself in the infant-juvenile age (before the age of 30) and the vast majority are of autoimmune origin. It is characterized by a defect in insulin secretion and constitutes 5–10% of all cases of diabetes. It is always a subsidiary of insulin treatment [2].

Diabetes type 2 (DM2): This is the most prevalent variety and is frequently linked to obesity or an increase in visceral fat. Ketoacidosis rarely develops spontaneously. The issue ranges from a predominant resistance to insulin, accompanied by a relative hormone deficiency, to a progressive malfunction in its secretion [1]. It is the most frequent form of DM2 since it represents between 90 and 95% of cases. It usually appears after the age of 40 and is associated with obesity, which is present in up to 80% of patients with type 2 DM. Its treatment requires diet and exercise alone or is associated with oral antidiabetics and/or insulin [2].

Gestational Diabetes Mellitus (GDM): Specifically, groups glucose intolerance detected for the first time during pregnancy. Hyperglycemia before twenty-four weeks of pregnancy is considered undiagnosed pre-existing diabetes [1]. It occurs in 1–14% of pregnant women and is associated with an increased risk of obstetric and perinatal complications [2].

Due to the interaction of numerous genetic variants and other environmental factors, diabetes mellitus type 2 (DM2) is a complex and multifaceted disorder characterized by chronic hyperglycemia. The prevalence of obesity and physical inactivity, together with the aging of the population, have all contributed to a significant rise in the number of people worldwide who have type 2 diabetes [3]. It is classified as one of the epidemics of the twenty-first century, both for its growing magnitude and for its negative impact on cardiovascular diseases [4].

DM2 is a heterogeneous disease of multifactorial etiology, in which insulin resistance and inadequate compensatory insulin secretion by pancreatic beta cells are combined; It manifests as chronic hyperglycemia, accompanied by carbohydrate, fat, and protein metabolism disorders. The susceptibility of this disease is determined by the combined effect of genetic and environmental factors [4].

Environment refers to all non-genetic factors that modulate the phenotype, which may include random environmental factors such as climate, geography, demographics, and socioeconomics; as well as the lifestyle that is made up of diet, smoking, alcoholism, and physical activity, which the individual can modify [4].

The disease is regarded as a polygenetic disturbance in which each genetic variety confers a partial and additive effect. Just 5–10% of cases of DM2 may be attributed to genetic defects; these cases include juvenile-onset diabetes, insulin-resistance syndromes, mitochondrial diabetes, and neonatal diabetes [5]. Examination of DM2 susceptibility genes may be useful for the prediction, prevention, and early treatment of the disease.

Through the implementation of genome-wide association studies (GWAS), the number of common genetic variants associated with DM2 has increased rapidly [6–12]. In addition, more than 40 genetic loci associated with DM2 have been identified; however, these loci have been identified primarily in European populations [13]. Still there are additional genetic factors to be discovered since the identified genetic regions only account for a small portion of the estimated heritability of DM2. The high economic cost and a large number of hypotheses in these studies are a limitation of GWAS [14]. Several research studies have examined cluster-based GWAS's viability and efficiency, with significant time and financial savings. [14–16]. In addition, whole genome sequencing across multiple samples in a population provides an unprecedented opportunity to comprehensively characterize polymorphic variants in the population [17].

Type 2 diabetes, as mentioned, is a complex illness brought on by numerous genetic and environmental factors; family-based and peer studies estimate that heredity ranges from 22 to 73%. Recent estimates placed the prevalence of DM2 in adults, adjusted for age, at 7.6% in European Americans, 14.9% in Afro-Americans, 4.3–8.2% in Asian Americans, and 10.9–15.6% in Hispanic Americans [18–21]. More than 40 genetic loci associated with DM2 have been identified, but so far, these locations have primarily been revealed through studies of people with European ancestry. The candidate gene association studies discovered a link between DM2 and nonsensical variants in PPARG (MIM 601487) and KCNJ11 (MIM 600937), which are targets for drugs to treat diabetes, and they implicated common genetic variants responsible for Mendelian forms of diabetes in DM2 [22–27].

The first genome-wide association studies (GWAS) for DM2 [6–9, 28] and fasting glucose [29] successfully identified multiple associated loci. And, through recent GWAS meta-analyses for DM2 [30] and quantitative glycemic characteristics [31], the number of loci associated with DM2 have significantly increased in European populations; the majority of these variants act via defects in the function of beta-cells rather than insulin action. In total, known variants associated with DM2 account for 10% of genetic variation [30, 32], therefore it is likely that more locations and independent factors increase the risk of the disease.

Few people outside of Europe are aware of the genetic factors that contribute to type 2 diabetes. A new locus (KCNQ1 [MIM 607542]) was discovered based on a GWAS in a Japanese population [33, 34] and was later discovered to have separate alleles in people of European ancestry [30]. Most recently, GWAS in Chinese populations [5, 35], Japanese [36], and south Asian [37] discovered additional DM2 loci that exceed genome-wide significance. To date, GWAS in African Americans has been underpowered to detect new loci [38].

In a recent multiethnic meta-analysis, three DM2 risk loci in Europe (GATAD2A/CILP2/PBX4, TH/INS, and SREBF1), one DM2 risk locus in Africa (HMGA2), and one DM2 risk locus in multiple ethnic groups (BCL2) were associated confirming that an allele-based gene score exists. Hence, the multiethnic GWAS of DM2 should result in the discovery of additional genes associated with diabetes that are relevant to numerous ethnic groups [13].

There are still additional genetic factors to be discovered since the identified genetic regions only account for a small portion of the estimated heritability of DM2. The high economic cost and a large number of hypotheses in these studies are a limitation of GWAS [14]. Several studies have looked at the viability and effectiveness of GWAS based on clusters, with considerable time and cost savings [14–16]. In addition, whole genome sequencing across multiple samples in a population provides an unprecedented opportunity to comprehensively characterize polymorphic variants in the population [17].

The purpose of this chapter was to perform an association study in the mitochondrial genome to identify Base-Pair (bp) genomic positions statistically associated with DM2. An alignment analysis enabled visualization and selection of a genomic region with allelic variability. Subsequently, a Principal Component Analysis (PCA) was used to visualize the complexity of the data; followed by a simple and multiple logistic regression analysis that allowed the discovery of base-pair positions associated with

DM2. Finally, an inspection of the mitochondrial genome annotation revealed 3 candidate genes to be associated with DM2.

2. Methodology

Next, the database used in this chapter will be explained, as well as the techniques used for the analysis of DNA sequences.

2.1 Database

We explored genetic variants of these type 2 diabetes-associated genes in different populations using genome-wide association analysis available in the Type 2 Diabetes Knowledge Portal database (http://www.type2diabetesgenetics.org/). The search criteria were: patients with DM2, considering a p-value <0.05 in the X2 test and an Odds Ratio > 1.0. Based on the results obtained, the variants were evaluated and identified in NCBI dbSNP (https://www.ncbi.nlm.nih.gov/snp/), and their registration was documented in ClinVar (https://www.ncbi.nlm.nih.gov/clinvar/). Related polymorphisms were explored in the UCSC Genome Browser (https://genome.ucsc.edu/) using the GRCh37/hg19 version and the change of polymorphisms as reference allele/effect and minimum allele frequency were compared with the information available in genomic databases. The allele frequency and the genotype frequency of the effect on heterozygotes and homozygotes were queried in the 1000 Genomes database using Ensembl (http://grch37.ensembl.org/index.html). Finally, samples of different tissues from patients with type 2 diabetes were analyzed with the Orange package (https://orange.biolab.si). To identify the differences in expression of this gene in different tissues, from GEO data sets (https://www.ncbi.nlm.nih.gov/gds) expression values of muscle, liver, and pancreas were obtained and the differences were analyzed by Mann Whitney U Test considering $p < 0.05$ significant.

To explore the prevalence and distribution of mitochondrial polymorphisms associated with DM2, the search for complete sequences of the mitochondrial chromosome (16,569 base pairs) was designed and minor fragments were considered; because most of the works on the subject are amplified for the control region (D-loop) with a size smaller than 1000 base pairs in the nucleotide database of the National Center for Biotechnology Information (NCBI) (https://www.ncbi.nlm.nih.gov/nucleotide) where the GenBank, the most extensive collection of genetics and genomics available, is located. Changes in the sequence were identified, as well as the insertion, deletion, and heteroplasmy sites. We also estimated the number of mitochondrial single nucleotide polymorphisms (mtSNPs), the average number of distinct nucleotides among populations, as well as the number of fixed differences, shared polymorphisms, and mono- and polymorphic mutations between populations. This made it easier to identify the polymorphisms that are most prevalent in the control region and the mtDNA codifying region.

The database is separated into two fasta files, the first file has 437 whole mitochondrial genome sequences from type 2 diabetic human patients and the second file has 73 from healthy individuals (each sequence is the most common or dominant whole mitochondrial chromosome in each individual, with 16,569 bases each, although there may be slight nucleotide variations between different individuals). The shortest sequence is 16,554 bases.

Both files were merged into a single file, with sick individuals placed before healthy ones, resulting in a total of 510 sequences. Once aligned, the sequences with the MEGA software have a length of 16,609 data. After this, a visual analysis of all the already aligned sequences was carried out, looking for the region that presented the greatest disadvantage, the region resulting from position 16,170–16,410, a total of 241 positions or data.

In addition, the nucleotides (Adenine (A), Cytosine (C), Guanine (G), Thymine (T)) were changed by numbers as follows, to perform a cluster analysis: A = 1, C = 2, G = 3 and T = 4, GAP(−) = 5. There were also other letters other than nucleotides such as R, Y, W, N. These letters were changed to the number 9. These last letters according to the nomenclature of the International Union of Pure and Applied Chemistry (IUPAC) correspond to:

- R = GA (purine)
- Y = TC (pyrimidine)
- W = AT (weak bonds)
- N = AGCT (any)

In addition, as extra information, we searched to which ethnic groups the people in the DNA sequences of the database belonged. Finding the following results: 239 sequences belong to Taiwanese people, 62 people are Indian, there are 6 Italians, 11 Chinese people, and 192 Japanese.

A total of 510 complete human mitochondrial genomes were used in this study. Of the total genomes, 437 were from people with DM2 and 73 from healthy people. The data was stored in a FASTA format file and the genomes were aligned using the CLUSTALW algorithm implemented in the MAFFT tool [39, 40]. The total length of the alignment was 16,610 nucleotides.

In the aligned genomes, an inspection was carried out to locate the region with the highest frequency, resulting in the detected region from position 16,170 to 16,410, with a length of 241 nucleotides. This region of the alignment was removed and the rest of the analysis was performed with these data.

2.2 Principal component analysis (PCA)

The main goal of principle component analysis (PCA) is to reduce the dimensionality of a set of data, which often consists of a large number of interrelated variables, while retaining all possible variation. This is accomplished by transforming a new group of variables known as the principal components (PCs), which are disassociated from one another and arranged so that the first few retain the most variation found in the total set of original variables [41].

Theoretically, PCA provides the best least squares transformation of a given set of data. In order to obtain the key components we provide a vector X^T of n dimensions, $X = [x^1, x^2, \ldots, x^n]^T$, whose mean vectors (M), and covariance (C) are described by $M = E(X) = [m_1, m_2, \ldots, m_n]^T$ and $C = E[(X - M)(X - M)^T]$. Then we calculate the eigenvalues $\lambda_1, \lambda_2, \ldots, \lambda_n$ and the eigenvectors $P_1, P_2, \ldots, P_n$; and order them according to their magnitude $\lambda_1 \geq \lambda_2 \geq \cdots \geq \lambda_n$. The d eigenvectors must be chosen to represent the n variables, $d < n$. Then $P_1, P_2, \ldots, P_d$ are known as principal components [41].

In order to apply PCA to the sequences, a transformation of the nucleotides was performed, from the ACGT format to the numerical format. Each nucleotide was assigned a value between 1 and 4 as follows: A = 1; C = 2; G = 3; and T = 4. In the same way, the blank spaces (GAP) = 5. The PCA analysis was applied to the resulting numerical matrix. The purpose of applying the PCA analysis was to analyze the structure of the data and look for possible clusters that differentiated the data from sick and healthy people.

2.3 Entropy analysis

Shannon's entropy theory, initially developed by Claude E. Shannon, is applied to measure the contrast between criteria and this information is used to make decisions. In this analysis, it is indicated that for all p_i within a probability distribution P, there is a measure H, which satisfies the following properties [42]:

1. H is a continuous positive function,
2. If all p_i is equal and $p_i = 1/n$, then H should be an increasing monotonic function of n; and,
3. For all, $n \geq 2$,

$$H(p_1, p_2, \dots, p_n) = h(p_1 + p_2, p_3 \dots, p_n) + (p_1 + p_2) H\left(\frac{p_1}{p_1 + p_2}, \frac{p_2}{p_1 + p_2}\right) \quad (1)$$

Shannon proved that the only function that satisfies these conditions is:

$$H_{Shannon} = -\sum_{i}^{n} p_i \log(p_i) \quad (2)$$

Where: $0 \leq p_i \leq 1; \sum_{i=1}^{n} p_i = 1$

2.4 Regression models

The objective of a linear regression model is to try to explain the relationship between a dependent variable (response variable) and a set of independent variables (explanatory variables) $X_1, \dots, X_n$. In a simple linear regression model, we try to explain the relationship between the response variable (Y) and a single explanatory variable (X). Using the regression techniques of a variable Y on a variable X, we look for a function that is a good approximation of a cloud of points (x_i, y_i), by means of a curve [43].

The variable dependency can be a univariate or multivariate regression. Univariate regression identifies the dependency between a single variable as represented in Eq. (2) [44].

$$Y = \alpha + \beta X + \varepsilon \quad (3)$$

Where y is a dependent variable, x is an independent variable with coefficient β (it is the slope of the line and indicates how Y changes when X increases by one unit), and α is a constant (it is the ordinate at the origin, the value which Y takes when X is 0), and ε a variable that includes a large set of factors, each of which influences the response only to a small magnitude, which we will call error. X and Y are random variables, so an exact linear relationship between them cannot be established [43]. While multivariate regression is to identify the dependence between several variables simultaneously, it is represented in Eq. (3) [44].

$$Y = \beta_0 + \beta_1 X_1 + \beta_2 X_2 + \cdots + \beta_p X_p + \varepsilon \tag{4}$$

Where ε is the error term, β_0 is the intercept, β_1-β_k are partial regression coefficients, for example, β_i when $1 \le i \le k$ represents the change in the mean response corresponding to a unit change in x_i when the other variables remain constant.

Regression models predict the outcome of the dependent variables from the independent variables. Importance is considered in regression analysis to handle more complicated problems [44]. The objective of multiple linear regression is to solve the set of coefficients $\Theta = \{\beta_0, \beta_1, \dots, \beta_k\}$ given the observations X and the objectives Y [45].

2.4.1 Linear regression

Linear regression is the most common predictive model to identify the relationship between variables. It can be simple linear or multiple linear regression. Linear regression is described in Eq. (4) [44].

$$y = x\beta + \varepsilon \tag{5}$$

In Eq. (4) y is the independent variable and can be a continuous or categorical value; x is a dependent variable that is always a continuous value. It analyzes a probability distribution and focuses mainly on conditional probability distribution with multivariate analysis [44].

2.4.2 Simple linear regression

The simple linear regression process that is depicted in **Figure 1** is a regression analysis that uses a single independent variable and is described in the Eq. (2). Similar to how correlation expands the relationship between two variables, simple linear regression distinguishes between dependent and independent variables; however, correlation does not do so [44].

2.4.3 Multiple linear regression

Multiple or Multivariate Linear Regression (MLR) depicted in **Figure 2** is the prediction process with more than one independent or predictor variable that is similar to multivariate analysis as described in Eq. (3) [44].

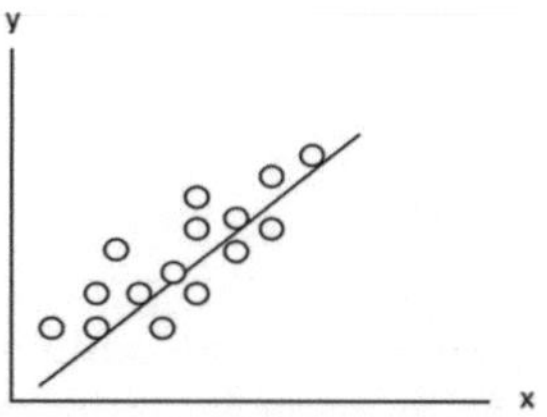

Figure 1.
Simple linear regression [44].

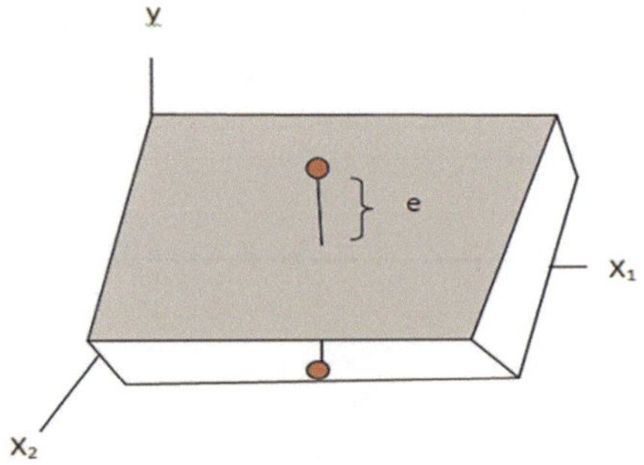

Figure 2.
Multiple linear regression [44].

A statistical technique known as multiple linear regression uses many explanatory variables to predict the outcome of a response variable. The multiple linear regression's goal is to model the relationship between the explanatory and response variables. The next model is a multiple linear regression model with k predictor variables, x_1, ..., x_k [45].

The MLR problem is frequently resolved using least squares. If each predictor variable x_1, x_2,. .., x_k has n observations, then x_{ij} represents the i-th observation of the j-th predictor variable x_j. For example, x_{31} represents the first value of the third observation. Specifically, Eq. (3) above can be expressed as [45]:

$$y_j = \beta_0 + \beta_1 X_{j1} + \beta_2 Xj_2 + \cdots + \beta_k X_{jk} + \varepsilon_j \tag{6}$$

Where $1 \le j \le n$, yj is the jth target value. The system of n equations can be represented as a design matrix as shown in Eq. (2), and describes the levels of the predictor variables acquired at each observations. All of the regression coefficients are contained in the vector β. The least squares estimates, which are stated below, are used to create the regression model β [45].

$$\hat{\beta} = (X^T X)^{-1} X^T y \tag{7}$$

Then the estimated value of y can be calculated as follows after obtaining $\hat{\beta}$ [45].

$$\begin{gathered} \hat{y} = X\hat{\beta} \\ \epsilon = y - \hat{y} \end{gathered} \tag{8}$$

The purpose of using regression data was to search for SNPs statistically associated with DM2.

2.5 Risk factors

A measure of the relationship between an exposure and a result is called an odds ratio (OR). The odds ratio (OR) shows the likelihood of an occurrence given a specific exposure in comparison to the likelihood of the outcome in the absence of that exposure. Case-control studies are the most frequent applications of odds ratios [46].

The odds ratio is used to compare the likelihood of an outcome (such a disease or disorder), because of exposure to a particular variable (e.g., health characteristic, item of medical history). The odds ratio can also be used to assess if a specific exposure represents a risk for a specific outcome and to assess the relative importance of several risk variables for that outcome [46].

- OR = 1 Outcome probabilities are unaffected by exposure.
- OR > 1 Exposure is linked to bigger odds of success.
- OR < 1 Exposure is linked to a reduced likelihood of success.

It is calculated using the 95% Confidence Interval (CI) to determine the accuracy of the OR. A high OR precision is indicated by a small CI, while a low OR precision is shown by a large CI. It is important to note that the 95% CI does not provide information about a measure's statistical significance, unlike the p-value. In reality, if the 95% CI does not overlap the null value (for instance, OR = 1), it is frequently regarded as a marker of statistical significance. Therefore, it would be incorrect to interpret a 95% CI OR that encompasses the null as showing that exposure and outcome are not related [46].

To define risk factors, each base-pair positions found to be significant in the association analysis (regression analysis) was inspected. The Odds Ratio (OR) calculation criteria and definition of Risk Factor, as described in [46] were applied. The statistical significance, OR value, and 95% confidence range for each variable were examined based on the findings. Then, each base-pair position that satisfied the subsequent requirements was declared as a risk factor:

1. If the base-pair position statistical significance (p-value) was less than 0.05;
2. The odds ratio (OR) was not equal to 1; and
3. The 95% confidence range for the odds ratio did not contain 1.

Hence, if a base-pair position satisfied these three criteria and its OR > 1, it is declared as a risk factor associated with a higher probability of diabetes. In the same way, if the variable met the three conditions, and its OR < 1, it is declared as a risk factor associated with a lower probability of diabetes.

3. Results

The complete mitochondrial genomes of the 510 patients were aligned with the MAFFT tool. The result of the alignment was visualized with the MEGA X software [47]. By visual inspection, one region with variability was observed, while the rest showed perfect alignment. **Figure 3a** shows a fragment of the region with variability, and **Figure 3b** shows a fragment of the region without variability.

The region between positions 16,170 and 16,410, with a length of 241 nucleotides, was chosen to perform the rest of the analysis.

To analyze the structure of the information and look for possible clusters that would differentiate the data from sick and healthy people, PCA was applied to the aligned region of high variability. To carry out this analysis, the statistical language R was used. **Figure 4** shows the graph of Principal Component 1 (PC1) against Principal

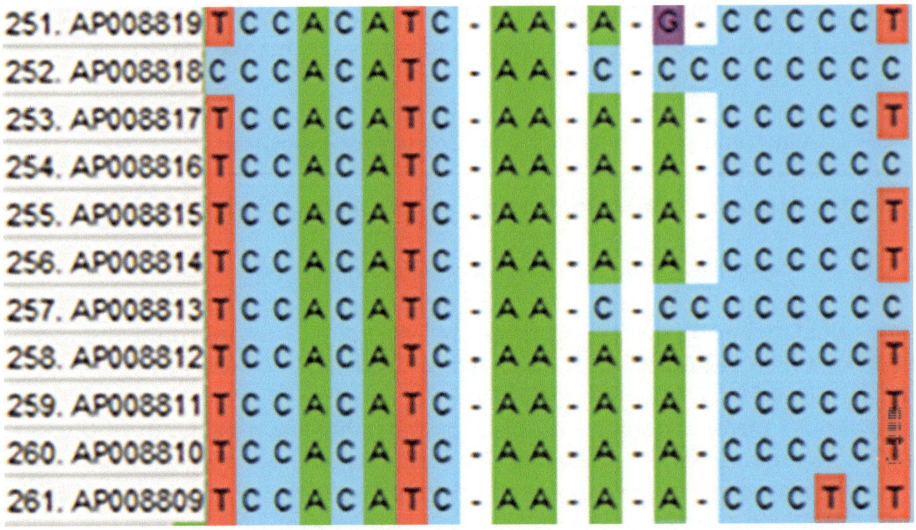

(a)

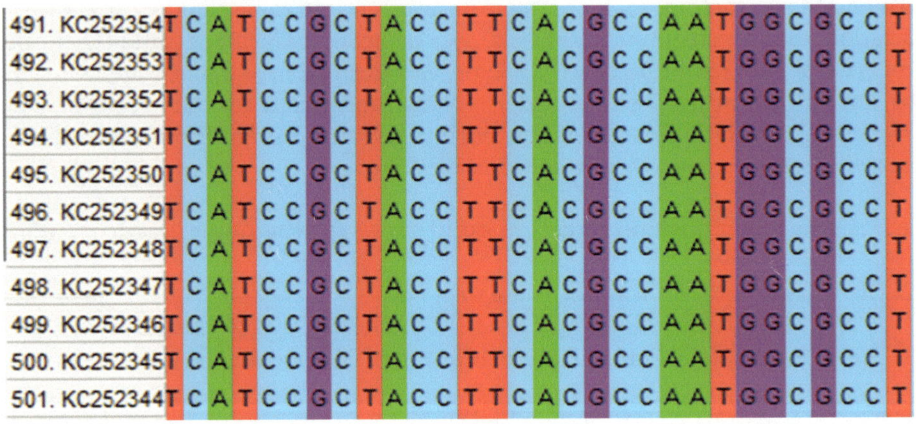

(b)

Figure 3.
Fragment of the alignment of the mitochondrial genomes of patients with DM2 and healthy. (a) Represents a region with variability and (b) represents a region with zero variability.

Component 2 (PC2). As we can see in the graph, the information appears mixed and there is no clear differentiation between the groups. This analysis shows us the complexity of the data.

The association analysis was performed in two steps, on the one hand, simple logistic regression was applied to each base-pair position (bp) of the variant region (241 base-pair positions), assigning a 1 to the dependent variable, for all healthy patients, and 0 to all patients with DM2. Those positions that were statistically significant (p-value <0.05) were selected. Subsequently, a multiple logistic regression was carried out grouping the positions that were significant in the simple regression. Those that were significant in the multiple regression were declared as positions associated with DM2. **Table 1** shows the positions that were significant both in the simple regression and in the multiple regression.

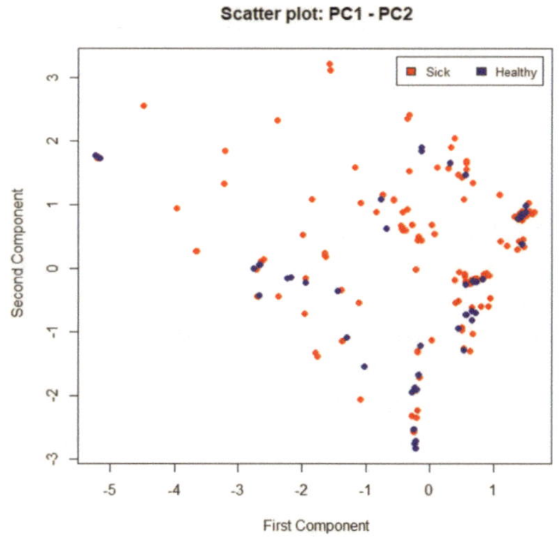

Figure 4.
Comparison graph between cases vs. controls.

Genomic position (BP)	Simple regression (P-value)	Multiple regression (P-value)	Associated with DM2
16,184	0.0038	0.0021	Yes
16,222	0.0384	0.6592	No
16,257	0.0289	0.1037	No
16,263	0.0415	0.6937	No
16,282	0.0033	0.0064	Yes
16,289	0.0426	0.4447	No
16,344	0.0038	0.0159	Yes
16,351	0.0438	0.1983	No

Table 1.
Simple and multiple regression results.

4. Discussion

As observed in **Table 1**, after multiple regression three positions were associated with DM2. The positions and their resulting p-values were: 16,184; 16,282, and 16,344 and 0.0021, 0.0064, and 0.0159, respectively. To locate the associated gene, the human mitochondrial genome annotation in the National Center for Biotechnology Information (NCBI) database (https://www.ncbi.nlm.nih.gov/) was inspected. Three genes were located within 3000 base pairs (bp) of the associated positions. These genes are: CYTB, which produces the Cytochrome B protein and contributes to the conversion of energy from food to cellular energy (Adenosine Triphosphate, ATP), the TRNP gene, which is the Proline tRNA, and the TRNT gene, which is the tRNA of Threonine. Especially the TRNT gene, was found to be associated with the maternal heritability of DM2 in Chinese families [48], and in another study carried out by Momiyama, et al., this gene was associated, as in our study, with the genomic position

16,184; and declared as one of the causes of left ventricular hypertrophy in patients with DM2 in Japanese families [49].

5. Conclusions

In this association study, 510 complete mitochondrial genomes were analyzed. Of the total genomes, 437 were from patients with DM2, and 73 from healthy patients. A genome-wide alignment allowed locating a variable region in its allelic content; a PCA analysis allowed us to visualize the complexity of the data, and a logistic regression analysis allowed us to find 3 base-pair positions associated with DM2. The associated positions were located within 3 k bp of three genes, one of which (TRNT gene) was reported by previous studies to be associated with DM2. Finally, this study adds new evidence of the association of genomic positions with DM2.

Acknowledgements

This work was developed within the Master's and Doctorate in Science and Engineering (MYDCI) program offered by the Autonomous University of Baja California. In addition, it was supported by a CONACYT scholarship.

Author details

Julio Alejandro Valdez[1,2*], Pedro Mayorga[2], Rafael Villa Angulo[1] and Carlos Villa Angulo[1]

1 Engineering Institute, Universidad Autónoma de Baja California, Mexicali, B.C., Mexico

2 Department of Electrical - Electronics, TecNM/ITMexicali, Mexicali, B.C., Mexico

*Address all correspondence to: a1174500@uabc.edu.mx

References

[1] de Rojas E, Molina R, Rodríguez C. Definición, clasificación y diagnóstico de la diabetes mellitus. Revista Venezolana de Endocrinología y Metabolismo. 2012; **10**(1):7-12. Available from: https://www.redalyc.org/articulo.oa?id=375540232003

[2] Mediavilla Bravo JJ. la diabetes mellitus tipo 2, Medicina Integral. Available from: https://www.elsevier.es/es-revista-medicina-integral-63-articulo-la-diabetes-mellitus-tipo-2-13025480

[3] Chen L, Magliano DJ, Zimmet PZ. The worldwide epidemiology of type 2 diabetes mellitus–present and future perspectives. Nature Reviews. Endocrinology. 2011;**8**(4):228-236. DOI: 10.1038/nrendo.2011.183

[4] Ofarrill LCL, Cuervo AM, Ferrer RL, Valdés MTL. Interacción genoma-ambiente en la diabetes mellitus tipo 2. Acta Médica del Centro. 2018;**12**(4). Available from: http://www.revactamedicacentro.sld.cu/index.php/amc/article/view/948

[5] Tsai FJ et al. A genome-wide association study identifies susceptibility variants for type 2 diabetes in Han Chinese. PLoS Genetics. 2010;**6**(2): e1000847. DOI: 10.1371/journal.pgen.1000847

[6] Scott LJ et al. A genome-wide association study of type 2 diabetes in Finns detects multiple susceptibility variants. Science. 2007;**316**(5829): 1341-1345. DOI: 10.1126/science.1142382

[7] Saxena R et al. Genome-wide association analysis identifies loci for type 2 diabetes and triglyceride levels. Science. 2007;**316**(5829):1331-1336. DOI: 10.1126/science.1142358

[8] Sladek R et al. A genome-wide association study identifies novel risk loci for type 2 diabetes. Nature. 2007; **445**(7130):881-885. DOI: 10.1038/nature05616

[9] Zeggini E et al. Replication of genome-wide association signals in UK samples reveals risk loci for type 2 diabetes. Science. 2007;**316**(5829): 1336-1341. DOI: 10.1126/science.1142364

[10] Burton PR et al. Genome-wide association study of 14,000 cases of seven common diseases and 3,000 shared controls. Nature. 2007; **447**(7145):661-678. DOI: 10.1038/nature05911

[11] Zeggini E et al. Meta-analysis of genome-wide association data and large-scale replication identifies additional susceptibility loci for type 2 diabetes. Nature Genetics. 2008;**40**(5):638-645. DOI: 10.1038/ng.120

[12] Gudmundsson J et al. Two variants on chromosome 17 confer prostate cancer risk, and the one in TCF2 protects against type 2 diabetes. Nature Genetics. 2007; **39**(8):977-983. DOI: 10.1038/ng2062

[13] Saxena R et al. Large-scale gene-centric meta-analysis across 39 studies identifies type 2 diabetes loci. American Journal of Human Genetics. 2012;**90**(3): 410-425. DOI: 10.1016/j.ajhg.2011.12.022

[14] Baum AE et al. A genome-wide association study implicates diacylglycerol kinase eta (DGKH) and several other genes in the etiology of bipolar disorder. Molecular Psychiatry. 2008;**13**(2):197-207. DOI: 10.1038/sj.mp.4002012

[15] Galvan A et al. Genome-wide association study in discordant sibships identifies multiple inherited

susceptibility alleles linked to lung cancer. Carcinogenesis. 2009;**31**(3): 462-465. DOI: 10.1093/carcin/bgp315

[16] Forstbauer LM et al. Genome-wide pooling approach identifies SPATA5 as a new susceptibility locus for alopecia areata. European Journal of Human Genetics. 2012;**20**(3):326-332. DOI: 10.1038/ejhg.2011.185

[17] Wong LP et al. Deep whole-genome sequencing of 100 southeast Asian Malays. American Journal of Human Genetics. 2013;**92**(1):52-66. DOI: 10.1016/j.ajhg.2012.12.005

[18] Cowie CC et al. Prevalence of diabetes and high risk for diabetes using A1C criteria in the U.S. population in 1988–2006. Diabetes Care. 2010;**33**(3): 562-568. DOI: 10.2337/dc09-1524

[19] Díaz-Apodaca BA, Ebrahim S, McCormack V, de Cosío FG, Ruiz-Holguín R. Prevalence of type 2 diabetes and impaired fasting glucose: Cross-sectional study of multiethnic adult population at the United States-Mexico border. Revista Panamericana de Salud Pública. 2010;**28**(3):174-181. DOI: 10.1590/s1020-49892010000900007

[20] Lee JW, Brancati FL, Yeh HC. Trends in the prevalence of type 2 diabetes in Asians versus whites: Results from the United States National Health Interview Survey, 1997-2008. Diabetes Care. 2011;**34**(2):353-357. DOI: 10.2337/dc10-0746

[21] Bowden DW et al. Review of the Diabetes Heart Study (DHS) family of studies: A comprehensively examined sample for genetic and epidemiological studies of type 2 diabetes and its complications. The Review of Diabetic Studies. 2010;7(3):188-201. DOI: 10.1900/rds.2010.7.188

[22] Altshuler D et al. The common PPARgamma Pro12Ala polymorphism is associated with decreased risk of type 2 diabetes. Nature Genetics. 2000;**26**(1): 76-80. DOI: 10.1038/79216

[23] Gloyn AL, Hashim Y, Ashcroft SJ, Ashfield R, Wiltshire S, Turner RC. Association studies of variants in promoter and coding regions of beta-cell ATP-sensitive K-channel genes SUR1 and Kir6.2 with Type 2 diabetes mellitus (UKPDS 53). Diabetic Medicine. 2001; **18**(3):206-212. DOI: 10.1046/j.1464-5491.2001.00449.x

[24] Sandhu MS et al. Common variants in WFS1 confer risk of type 2 diabetes. Nature Genetics. 2007;**39**(8):951-953. DOI: 10.1038/ng2067

[25] Winckler W et al. Evaluation of common variants in the six known maturity-onset diabetes of the young (MODY) genes for association with type 2 diabetes. Diabetes. 2007;**56**(3): 685-693. DOI: 10.2337/db06-0202

[26] Winckler W et al. Association of common variation in the HNF1alpha gene region with risk of type 2 diabetes. Diabetes. 2005;**54**(8):2336-2342. DOI: 10.2337/diabetes.54.8.2336

[27] Winckler W et al. Association testing of variants in the hepatocyte nuclear factor 4alpha gene with risk of type 2 diabetes in 7,883 people. Diabetes. 2005; **54**(3):886-892. DOI: 10.2337/diabetes.54.3.886

[28] Steinthorsdottir V et al. A variant in CDKAL1 influences insulin response and risk of type 2 diabetes. Nature Genetics. 2007;**39**(6):770-775. DOI: 10.1038/ng2043

[29] Bouatia-Naji N et al. A polymorphism within the G6PC2 gene is associated with fasting plasma glucosc

levels. Science. 2008;**320**(5879): 1085-1088. DOI: 10.1126/science.1156849

[30] Voight BF et al. Twelve type 2 diabetes susceptibility loci identified through large-scale association analysis. Nature Genetics. 2010;**42**(7):579-589. DOI: 10.1038/ng.609

[31] Dupuis J et al. New genetic loci implicated in fasting glucose homeostasis and their impact on type 2 diabetes risk. Nature Genetics. 2010;**42**(2):105-116. DOI: 10.1038/ng.520

[32] So HC, Gui AH, Cherny SS, Sham PC. Evaluating the heritability explained by known susceptibility variants: A survey of ten complex diseases. Genetic Epidemiology. 2011; **35**(5):310-317. DOI: 10.1002/gepi.20579

[33] Yasuda K et al. Variants in KCNQ1 are associated with susceptibility to type 2 diabetes mellitus. Nature Genetics. 2008;**40**(9):1092-1097. DOI: 10.1038/ng.207

[34] Unoki H et al. SNPs in KCNQ1 are associated with susceptibility to type 2 diabetes in East Asian and European populations. Nature Genetics. 2008; **40**(9):1098-1102. DOI: 10.1038/ng.208

[35] Shu XO et al. Identification of new genetic risk variants for type 2 diabetes. PLoS Genetics. 2010;**6**(9):e1001127. DOI: 10.1371/journal.pgen.1001127

[36] Yamauchi T et al. A genome-wide association study in the Japanese population identifies susceptibility loci for type 2 diabetes at UBE2E2 and C2CD4A-C2CD4B. Nature Genetics. 2010;**42**(10):864-868. DOI: 10.1038/ng.660

[37] Kooner JS et al. Genome-wide association study in individuals of South Asian ancestry identifies six new type 2 diabetes susceptibility loci. Nature Genetics. 2011;**43**(10):984-989. DOI: 10.1038/ng.921

[38] Lettre G et al. Genome-wide association study of coronary heart disease and its risk factors in 8,090 African Americans: The NHLBI CARe Project. PLoS Genetics. 2011;**7**(2): e1001300. DOI: 10.1371/journal.pgen.1001300

[39] Kuraku S, Zmasek CM, Nishimura O, Katoh K. aLeaves facilitates on-demand exploration of metazoan gene family trees on MAFFT sequence alignment server with enhanced interactivity. Nucleic Acids Research. 2013;**41**(W1):W22-W28. DOI: 10.1093/nar/gkt389

[40] Katoh K, Rozewicki J, Yamada KD. MAFFT online service: Multiple sequence alignment, interactive sequence choice and visualization. Briefings in Bioinformatics. 2017;**20**(4): 1160-1166. DOI: 10.1093/bib/bbx108

[41] Mateos-Valenzuela AG, González-Macías ME, Ahumada-Valdez S, Villa-Angulo C, Villa-Angulo R. Risk factors and association of body composition components for lumbar disc herniation in Northwest, Mexico. Scientific Reports. 2020;**10**(1):18479. DOI: 10.1038/s41598-020-75540-5

[42] Delgado A, Huamani A, Brillitt B. Applying Shannon Entropy to Analise Health System Level by departments in Peru. In: in 2018 IEEE XXV International Conference on Electronics, Electrical Engineering and Computing (INTERCON), 8-10 Aug. 2018. 2018. pp. 1-4. DOI: 10.1109/INTERCON.2018.8526435

[43] Limeres CC. Regresión lineal simple. http://eio.usc.es/eipc1/BASE/

BASEMASTER/FORMULARIOS-PHP-DPTO/MATERIALES/Mat_50140116_Regr_%20simple_2011_12.pdf [Accessed: 19 Enero 2022]

[44] Kavitha S, Varuna S, Ramya R. A comparative analysis on linear regression and support vector regression. In: in 2016 Online International Conference on Green Engineering and Technologies (IC-GET), 19-19 Nov. 2016. 2016. pp. 1-5. DOI: 10.1109/GET.2016.7916627

[45] Zhang Z, Li Y, Li L, Li Z, Liu S. Multiple linear regression for high efficiency video intra coding. In: ICASSP 2019 - 2019 IEEE International Conference on Acoustics, Speech and Signal Processing (ICASSP), 12-17 May 2019. 2019. pp. 1832-1836. DOI: 10.1109/ICASSP.2019.8682358

[46] Szumilas M. Explaining odds ratios. Journal of Canadian Academy of Child and Adolescent Psychiatry. 2010;**19**(3): 227-229. Available from: https://pubmed.ncbi.nlm.nih.gov/20842279

[47] Kumar S, Stecher G, Li M, Knyaz C, Tamura K. MEGA X: Molecular evolutionary genetics analysis across computing platforms. Molecular Biology and Evolution. 2018;**35**(6):1547-1549. DOI: 10.1093/molbev/msy096

[48] Li K, Wu L, Liu J, Lin W, Qi Q, Zhao T. Maternally inherited diabetes mellitus associated with a novel m.15897G>A mutation in mitochondrial tRNA(Thr) gene. Journal Diabetes Research. 2020;**2020**:2057187. DOI: 10.1155/2020/2057187

[49] Momiyama Y et al. A mitochondrial DNA variant associated with left ventricular hypertrophy in diabetes. Biochemical and Biophysical Research Communications. 2003;**312**(3):858-864. DOI: 10.1016/j.bbrc.2003.10.195

Chapter 6

Multiple Genetic Polymorphisms within microRNA Targets and Homologous microRNA-Binding Sites: Two More Factors Influencing microRNA-Mediated Regulation of Gene Expression

Miruna Giurgiu, Robert Kaltenbach, Franziska Ahrend, Summer Weeks, Holly Clifton, Martin Bouldo, Vitaly Voloshin, Jiling Zhong, Siegfried Harden and Alexander Kofman

Abstract

miRNA-mRNA interaction depends on multiple factors such as 3'UTR isoforms, the cell and tissue-specific expression levels of RNA-binding proteins, the sequence context around the mRNA target site, and other mechanisms. Genetic polymorphisms within miRNAs and their target sites appear to be among the most important ones because they influence the mode and outcome of miRNA-mRNA interaction universally and irreversibly. SNP disruption of miRNAs and their binding sites, as well as conformational changes preventing the access of the miRNA to its target site, are adopted as the most credible mechanistic explanations of SNP-mediated effects. The occurrence of multiple SNPs within the same miRNA-binding site implies their combinatorial mode of action. The presence of the repetitive (homologous) binding sites for the same miRNA on its mRNA target may both enhance the miRNA targeting and provide for the backup target site instead of the one disrupted by SNP, thus rescuing the miRNA functionality. While being underexplored, the multiple genetic polymorphisms within the miRNA-binding sites, as well as homologous miRNA-binding sites, may be considered as additional factors influencing miRNA-mediated regulation of gene expression.

Keywords: microRNA, mRNA, target, single-nucleotide polymorphism, homologous microRNA-binding sites

1. Introduction

Since the beginning of the 2000s, miRNAs stay in the focus of every aspect of medical and biomedical research [1, 2]. miRNAs are involved in a wide range of biological processes such as cell differentiation [3–7] and reprogramming [8], cellular senescence [9] and cell death [10–16], tissue and organ development [5, 17–20] and regeneration [21–26], cell signaling [27, 28], oxidative stress and metabolism [3, 29–37], mitochondrial dysfunctions [22, 38, 39], hormonal regulation [16, 40, 41] and adaptive responses [17], brain function [42, 43], inflammation [31, 44–46], immune response and the effects of the microbiome [46–53], viral infections and latency [43, 54–61], DNA damage and repair [22, 62–67], genomic balance [68] and genomic instability [65, 69], etc.

miRNAs deregulation promotes the biological processes resulting in various human diseases [41, 70–84] including oncogenesis [6, 7, 35, 37, 54, 59, 84–120]. miRNAs are recognized as important factors that are involved in tumor cell invasiveness [121], metastases [90, 121–129], cancer cachexia [130], drug resistance [131–136], sensitivity to chemotherapy [65, 137, 138], and radiosensitivity [138–140]. miRNAs are actively explored as both therapeutic agents [32, 112, 141–164] and targets [28, 42, 161, 165–174]. A growing number of reports point to the usefulness of miRNAs, and specifically circulating miRNAs [175, 176] as diagnostic and prognostic biomarkers [51, 70, 71, 73, 110, 112, 155, 160, 161, 174, 177–223] related to clinical and forensic [177, 224] studies, as well as normal physiological conditions, such as, diet [225] and increased physical activity [226, 227].

Deep sequencing and computational approaches indicate that miRNA genes comprise about 0.5–1% of the predicted genes in animals and humans [228]. One mRNA can be targeted by hundreds of miRNAs [84, 229], and a single miRNA can potentially recognize hundreds of different target transcripts [229, 230], which often share pathways to ensure their impact [52]. Therefore, miRNAs may be considered as global regulators of gene expression [231] with the vast *targetome* [52, 174] regulating about 90% of all protein-coding genes [232–234] and *interactome*-complex networking and cooperation between themselves [235, 236], as well as other regulatory molecules, in particular, transcription factors and noncoding RNAs [237–239].

Identification and characterization of the factors influencing miRNA functionality are essential to elucidate the mechanisms of miRNA activity and to explain and predict the effects of miRNAs for clinical applications. Genetic variations of both miRNAs and their target sites may have a significant impact on the efficacy of miRNA targeting. It is growingly recognized that naturally occurring variations in miRNAs and their target genes contribute to phenotypic complexity and may be associated with human pathologies [240, 241], including cancer [231, 242–244]. We present information about two underexplored SNP-related mechanisms influencing miRNA-mRNA interaction: the multiple SNPs within the single miRNA target site and the multiple (homologous) target sites for the same miRNA.

2. miRNA properties and biogenesis

miRNAs belong to the category of small noncoding RNA molecules [2, 245] ranging between 19 and 24 nucleotides in length [229]. miRNAs have been detected in all biological species, including viruses [55–60, 246]. Of note, bacteria do have short

RNA sequences with the miRNA-similar functions, yet they are not considered as true miRNAs [247]. Bioinformatics analysis indicates the existence of miRNAs derived from transposable elements [248, 249]. Although miRNAs are usually located in the cytoplasm, several studies have detected a regulatory role of miRNAs in other cell compartments such as mitochondria *(mitomiRs)*, nucleus, endoplasmic reticulum, granules, and P-bodies [38, 232]. Nuclear miRNAs participate in the regulation of miRNA biogenesis and function [250].

Circulating miRNA may be released by different types of cells and delivered into recipient cells for functional purposes, acting as cell-to-cell signaling mediators [251]. miRNA may be freed from dying cells or selectively sorted into the secreted small vesicles called exosomes [158, 229, 252–255]. Such exosomal miRNAs are regarded as promising clinical biomarkers (see above). miRNA can undergo both vertical and horizontal transmissions among distinct species, remarkably through feeding between plants and animals [229, 245].

The biogenesis of canonical animal miRNAs starts with the transcription of the pri-miRNA by RNA Pol II [256, 257]. The pri-miRNA has the length of about several thousands of nucleotides [258], the 7-methylguanosine, and 3′- poly(A) tail [259]. The pri-miRNA forms a stem-loop structure, which is recognized and cleaved by *Drosha* (endonuclease RNAse III) and two molecules of its partner protein, DGCR8 (*DiGeorge syndrome critical region 8*, named *Pasha* in flies and nematodes), which results in the formation of pre-miRNA. The pre-miRNA is further processed by the *Dicer* (another endonuclease RNAse III). The Dicer's partner protein TRBP (transactivation response element RNA-binding protein) is essential for miRNA processing in flies. The resulting short RNA duplex is associated with one of the AGO proteins to form the RISC, which is essential in destabilizing the target mRNA molecule or inhibiting its translation (for detailed review of miRNA biogenesis see [229, 260–263]). Noncanonical miRNAs are produced without *Drosha*- or *Dicer*-catalyzed cleavage [84, 261, 264]. Each step of miRNA biogenesis is tightly regulated [260, 263]. Like all RNA molecules, miRNAs undergo posttranscriptional modifications, which may affect their properties [234, 260, 262, 265, 266]. Recent studies point to the existence of sequence variants in miRNAs called *isomiRs*, which differ from the annotated miRNAs by altered sequences due to the various posttranscriptional modifications [267, 268].

The core miRNA function is posttranscriptional regulation of the expression of targeted genes. It is achieved through miRNA-mediated RNA silencing [269]. miRNAs, as a part of the RISC, trigger the various forms of the translational repression of the target mRNA *via* blocking its ribosomal loading and initiation of translation, inhibiting translation elongation, and causing protein degradation [234, 261, 262]. It may also cause mRNA deadenylation and decay [234], which is typical for plants, whereas translational repression is more characteristic for animal cells [261, 270, 271]. In contrary to mRNA degradation, translational repression is reversible and can be employed for rapid response to internal or external cues [271].

While target repression is universally recognized as the dominant mode of miRNA action, some reports indicate that miRNAs can stabilize the targeted transcript [234, 272–276]. MiRNAs can also regulate gene expression both indirectly by targeting the mRNAs of transcription factors and directly after being transported back to the nucleus and binding to the complementary sequence in the promoter [232, 234, 250, 262, 276]. miRNAs can also perform other noncanonical functions acting as ligands for toll-like receptors [277].

3. MiRNA target recognition

miRNAs bind to the complementary sequences of the host mRNA as a part of the RISC with its core component *Ago* proteins [229]. In plants, miRNAs show perfect or near-perfect complementarity to their target mRNA. In mammalian cells, as well as in other eukaryotes, miRNAs bind to mRNA sequences with imperfect complementarity [229, 278]. The 5′-end of miRNA exhibits the so-called *seed* sequence (**Figure 1**), which is complementary to six or more nucleotides at the 3'UTR of mRNA [2, 233, 262, 279].

Canonical miRNA-binding sites are positioned in the 3'UTR of the targeted mRNA sequence and classified upon the extent and location of matching miRNA nucleotides: 6mers perfectly pair to nucleotides 2–7 on the 5′-end of miRNA, 7-merA1 and 7mer8—additional pairing with miRNA nucleotide 1 or 8 respectively, and 8mer sites match miRNA nucleotides 1–8 [261]. The efficacy of these sites is usually augmented by additional pairing at the 3′ end of miRNAs. In addition to the binding site at the miRNA 5′ end, pairing around the binding sites contributes to the targeting efficacy [233, 261, 280]. The so-called 3′-supplementary sites with atypical elaboration of the 6mer, 7mer, and 8mer sites (≥3–4 pairs), and 3′-compensatory sites stretching for more than 4–5 pairs, while the seed region is mismatched are called atypical sites [84, 233, 261]. The supplementary pairing is believed to enhance the miRNA-mRNA interaction with the greater specificity. However, in animal cells, the high complementarity beyond the seed may cause TDMD [281], the common mechanism of destabilization of the short RNA molecules including miRNAs, which is observed in many diseases [84]. In general, the miRNA-mRNA complementarity with nucleotides at 3′ end or with nucleotides in the center (centered pairing) in the absence of the perfectly matched seed nucleotides is believed to be much less effective in regulating gene expression [84, 233, 261, 282–284]. The noncanonical seed-matching sites may also be in the protein-coding regions of the target mRNAs and even in the 5'UTR mRNA sequence [84, 232, 233, 285, 286].

4. Factors influencing miRNA targeting

miRNA-mRNA interaction depends on a range of factors. The posttranscriptional modification resulting in miRNA methylation causes structural changes that affect AGO binding [232]. miRNA sponges [287–289] bind miRNAs, thus preventing them from interacting with mRNA targets. miRNAs competition and cross talk with RNA-binding proteins [84, 232, 233, 236, 270], whose expression levels may depend on the cellular context and vary between the tissues [261, 290], also add up to the spectrum of various regulatory outcomes beside the dominating mechanism of

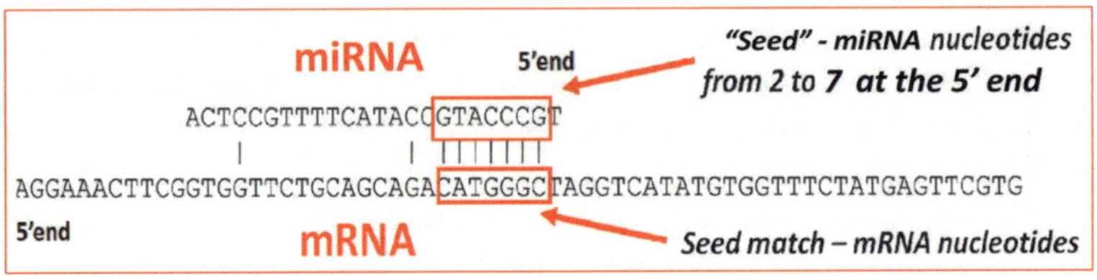

Figure 1.
miRNA-mRNA interaction in animal cells (developed by F. Ahrend).

miRNA targeting with moderate target repression [84]. miRNA targeting is known to depend on 3'UTR isoforms: 3'UTR size, alternative cleavage, and the location of target sequences within AU-rich regions [233, 261, 282, 284, 290–292].

The extent of complementary identity between miRNA and its mRNA targets appears to be one of the crucial factors, which determine the mode of target repression. The target cleavage by RISC is typical for plants, where the pairwise alignment is characterized by high complementarity. In opposite, the limited seed matching in animals often leads to translational inhibition [270]. It has been suggested that the above-mentioned noncanonical stabilization of the targeted transcript by miRNAs may depend on the complementarity between the miRNA and its target mRNA [275]. On the other hand, the extensive base-pairing may also cause TDMD [84, 293]. In this regard, genetic polymorphisms within miRNAs and their target sites appear to be among the most important mechanisms that ubiquitously and irreversibly define the mode and outcome of miRNA-mRNA interaction.

5. miRNAs and genetic polymorphisms

SNPs are genetic variations (nucleotide substitutions and indels) in DNA sequence [294, 295]. SNPs within the mRNA noncoding regions do not change the protein sequence, but they may significantly alter gene expression by affecting transcription, RNA processing, translation, and interaction of mRNA with noncoding regulatory RNAs [296]. While most SNPs are likely to be functionally neutral, some of them may represent causative links with human diseases [297]. SNPs in miRNA-precursor flanking regions, promoters of miRNA-encoding genes, and in the genes involved in miRNA biogenesis (transcription and RNA processing) may result in higher or lower miRNA expression profiles [298–302], which, in turn, represent the unique disease-specific signatures that can identify cancer types [231, 240, 243, 303]. SNPs within mature miRNA sequences, specifically in their seed regions, dissimilarities in mature miRNAs due to variable cleavage sites for *Drosha* and *Dicer* [243, 301, 302, 304], and polymorphisms within or nearby miRNA-binding sites on genes targets, may enhance or weaken the interaction between miRNA and its target transcripts and contribute to phenotypic variations and disease susceptibility [230, 240, 241, 267, 294, 305–321].

Within the mRNA targets, SNPs can either create or destroy miRNA-binding sites. It may impair miRNA ability to target oncogenes and, in opposite, render tumor-suppressor genes susceptible to miRNA-mediated inhibition (**Figure 2**). SNPs located both within and beyond the seed-target regions [322] can affect miRNA-mRNA interaction by destabilizing the mRNA target molecule, slowing down its ability for ribosomal loading, and altering the secondary RNA structure, which in turn, may have an impact on the availability of miRNA-binding sites [284, 290, 323]. Disruption or creation of miRNA-binding sites is recognized as one of the most credible mechanistic explanations of SNP-mediated effects [240, 241].

6. Multiple SNPs within miRNA-binding sites

A considerable number of genome-wide association studies demonstrated links between the SNPs within the miRNA target sites and predisposition to various human diseases. The attention was mostly devoted to SNPs, whose population frequency was

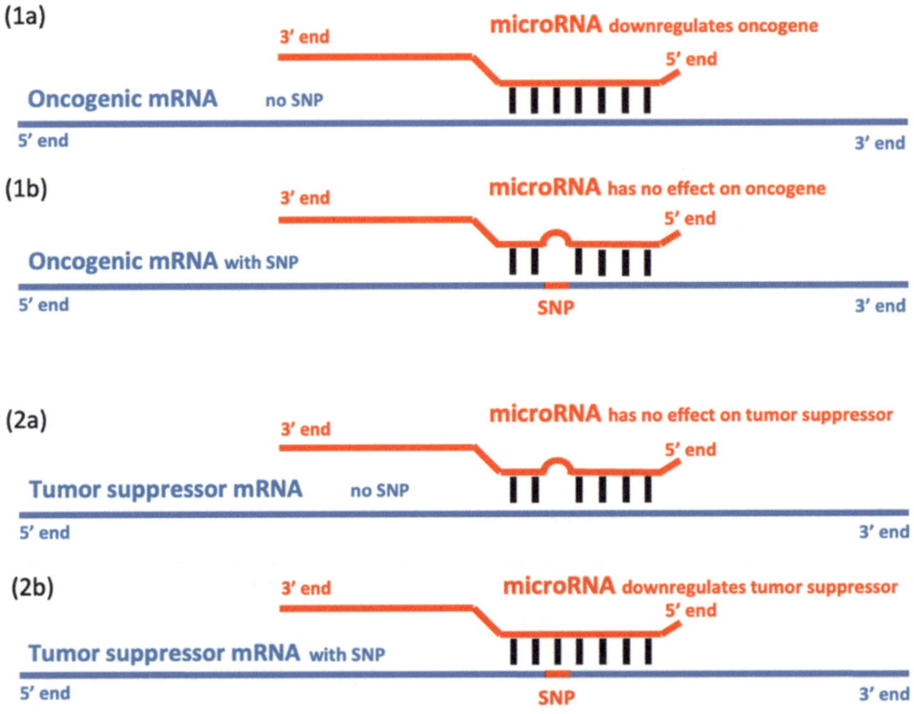

Figure 2.
The hypothetical effects of SNP on miRNA-mRNA interaction and cancer development. ***1a****—miRNA suppresses the oncogenic mRNA,* ***1b****—SNP disrupts the target site resulting in oncogene overexpression,* ***2a****—miRNA does not control the expression of the tumor suppressor gene, and* ***2b****—SNP creates the miRNA-binding site resulting in the inhibition of the tumor suppressor gene expression (developed by H. Clifton and F. Ahrend).*

well above 1%. Still, miRNA target sites may harbor multiple SNPs, although most of them are of low population frequency [324]. Even though the probability of their coincidental occurrence within the site is low, the frequency of some SNPs may vary significantly due to their ethnic disparities [325]. It has been suggested that, at least for some genes, it is not a separately taken SNP, but the combinatorial effect of several SNPs that may determine the outcome and efficacy of miRNA-mediated regulation of gene expression [326]. The location of SNPs (seed-corresponding region, centered, or 3′ end position) should also be taken into consideration.

Thus, the presence of additional, even rare (low frequency) SNPs within miRNA target sites, may modify (weaken or enhance) the effects of the SNPs, which occur at higher frequencies. Consequently, multiple SNPs may increase the probability of the site disruption, but may be also neutral, and even enhancing (**Figure 3**). This situation is less typical for the protein-coding regions due to the selection pressure resulting in the fewer SNPs. In the case of the overlapping miRNA-binding sites, the outcome is less clear due to the co-targeting effect—additive repression by more than one miRNA [84, 232]. Beside this, the increased complementarity between miRNA and its target in animals may result in miRNA decay (see above about TDMD).

SNPs within the single target site may be independent (**Figure 4**) and mutually exclusive if they are overlapping (**Figure 5**). Of note, SNP frequencies in the databases reflect the experimental NextGen sequencing data, and the possible SNP overlap is not

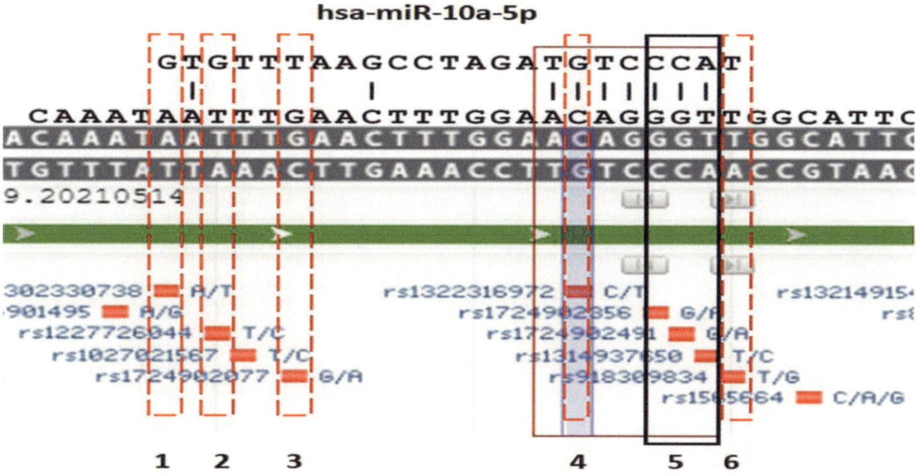

Figure 3.
The examples of disrupting, enhancing, and neutral SNPs within a single target site. The image shows the fragment of the human PDGFRA DNA sequence corresponding to the mRNA 3'UTR region, the miRNA-10a-5P binding site. **1**—*Enhancing through the formation of the G·U wobble base pair (rs1302330738),* **2**—*Neutral switch from the G·U wobble base pair to the classical Watson-Crick G·C bond (rs1227726044),* **3**—*Neutral switch from the G·U wobble base pair to the classical Watson-Crick A-T bond,* **4**—**N***eutral switch from the classical Watson-Crick G·C bond to the G·U wobble base pair (rs1322316972),* **5**—*Disrupting (rs1724902856, rs1724902491, rs1314937650), and* **6**—*Enhancing through the formation of the G·U wobble base pair (rs918309834). Note: The G·U wobble base pair has comparable thermodynamic stability to that of Watson-Crick base pairs and is known as a fundamental unit of RNA secondary structure in every class of RNA in organisms of all three phylogenetic domains [327]. However, it is still under the discussion whether G·U is characteristic for the canonical mRNA seed match [328] (developed by M. Giurgiu).*

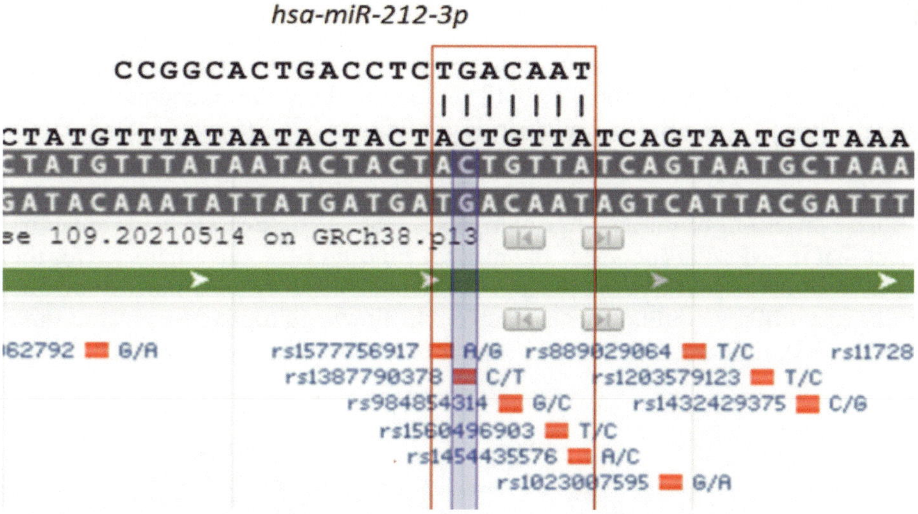

Figure 4.
The fragment of the human PDGFRA DNA sequence corresponding to the mRNA 3'UTR region, the miRNA-212-3P binding site. SNPs rs1577756917, rs1387790378, rs984854314, rs1560496903, and rs1454435576 represent independent events (developed by M. Giurgiu).

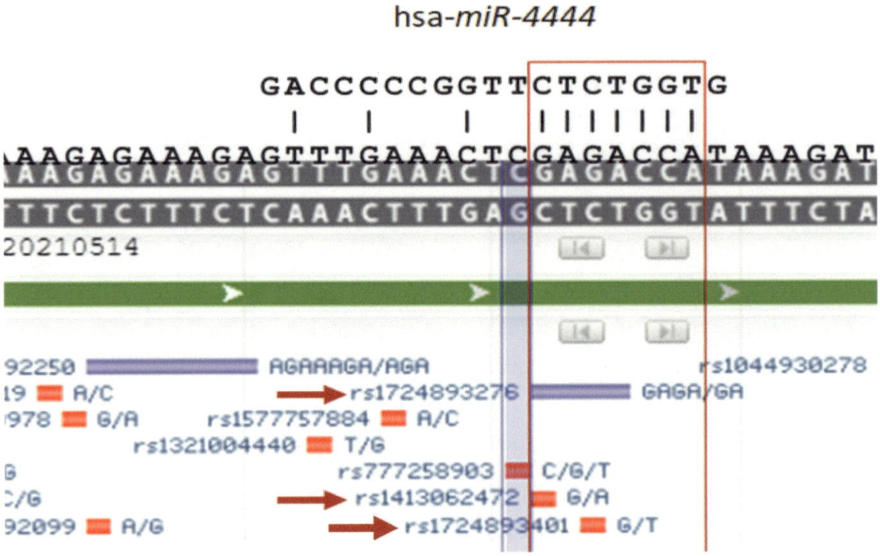

Figure 5.
The fragment of the human PDGFRA DNA sequence corresponding to the mRNA 3'UTR region, the miRNA-4444 binding site. SNPs rs1724893276 and rs1413062472, as well as rs1724893401, represent mutually exclusive events. rs1724893276 is an indel, and in the case of the deletion, neither rs1413062472 nor s172489 3401 (nucleotide substitutions) will be present (developed by M. Giurgiu).

taken into consideration. If mutually exclusive SNPs are represented by the overlapping indel and nucleotide substitution, the latter one is "shadowed" by the indel (**Figure 5**), and its real population frequency may be higher than that presented in the database.

7. Homologous miRNA-binding sites

It has been reported that some mRNAs may have more than one binding (seed-matching) site for the miRNAs [329–336]. These sites have the analogous seed-binding motif, but the different nucleotide content outside (**Figure 6**). Such multiple sites are proposed to act synergistically [330, 333–335, 337, 338], which, in turn, may depend on the distance between the sites [330, 338], as well as the activity and structural variations of miRNA-Ago complexes [336, 339]. It would be rational to suggest that if SNPs disrupt one of the miRNA-binding sites, the other sites for the same miRNA could preserve the miRNA-mediated control of gene expression. Such homologous miRNA-binding sites are present in almost all human genes, and their numbers are significant and directly proportional to the length of the 3'UTR [340].

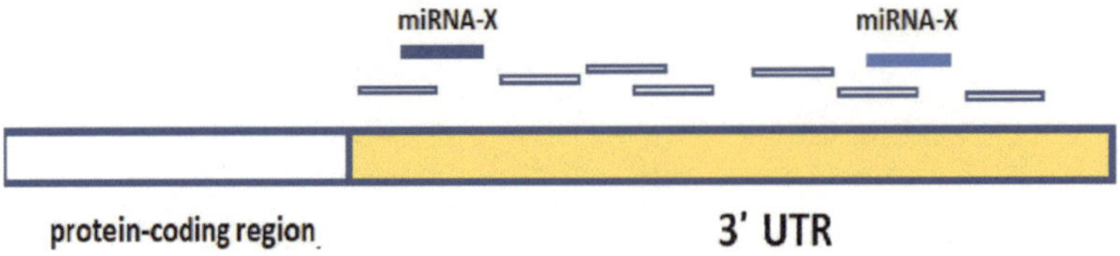

Figure 6.
The repeated (homologous) target sites for miRNA-X (developed by A. Kofman).

8. Bioinformatic tools

Whereas miRNAs and SNPs are both subjects of intense studies, there are few resources that allow to investigate the potential effects of SNPs on miRNA functionality.

PolymiRTS (Polymorphism in miRNAs and their Target Sites) is a database of naturally occurring DNA variations in miRNAs seed regions and miRNAs target genes [341]. *PolymiRTS* database was created by scanning 3'UTRs of mRNAs in humans and mice for SNPs in miRNA target sites. The database contains the results of genome-wide association studies about the links between the specific SNPs and human diseases. The PolymiRTS database also includes polymorphisms in target sites that have been supported by a variety of experimental methods and polymorphisms in miRNA seed regions.

MiRSNP is another database of human polymorphisms altering miRNA-binding sites [342]. It is a collection of 414,510 SNPs that are predicted to be able to affect (decrease/break or enhance/create) miRNA-mRNA binding.

MiRdSNP [343] is a database supplemented with the tools allowing the proximity searches between miRNA target sites and disease-associated SNPs (dSNPs) by gene name, miRbase ID, target prediction algorithm, disease, and any nucleotide distance between dSNPs and miRNA target sites. The web interface displays detailed sequence views showing the relationship among dSNPs, miRNA target sites, and SNPs. An interactive visualization tool shows the chromosomal distribution of dSNPs, miRNA target sites from TargetScan, and SNPs. The limitation of *MiRdSNP* is that the database contains only manually curated dSNPs on the 3'UTRs of human genes from available publications in PubMed, and the SNPs are only those that are disease-associated.

MSSD [344] is another manually curated database that provides comprehensive experimentally supported associations among miRNAs, SNPs, and human diseases. It has the same limitations as *MiRdSNP*.

miRNAhrend (https://franziskaahrend.pythonanywhere.com/) is the online-available package (beta version), which allows users to predict miRNA-target sites in any transcript upon the presence of seed-matching 6mers, which is verified by the complementarity-based alignment. The tool allows to assess how the presence of SNPs changes ("disappear/emerge") the landscape of the target sites and their corresponding miRNAs, identify the homologous miRNA-binding sites, and track down the exact position of the sites within any studied RNA sequence.

9. Conclusions

The ability of miRNAs to regulate the variety of genes in all tissues, as well as their presence in all biological species and their promise for clinical applications as novel diagnostics and therapeutics prompt further exploration of the factors influencing miRNA activity and functions. It is possible to assume that the presence of SNPs within miRNAs and their binding sites represents a powerful mechanism that can influence all biological processes in which miRNAs are involved. However, the growing number of reports indicates that the presence of SNPs cannot be interpreted as unalterable situation defining the biological outcome of miRNA targeting. The combinatorial effect of multiple SNPs located within the same target site, as well as beyond the target site, yet within the region where additional complementary matches between miRNA and its target can be formed, the circumstances where some SNPs may be mutually exclusive, and the homologous sites, which hypothetically can be utilized by miRNAs as an alternative target if the other sites are disrupted or become inaccessible, all these additional factors related to miRNA functionalities are anticipated of being studied with the development of the appropriate biocomputing tools and mathematical models.

Abbreviations

miRNA	microRNA
mRNA	messenger RNA
SNP	single-nucleotide polymorphism
Pol	polymerase
pri-miRNA	primary miRNA transcript
pre-miRNA	precursor miRNA
AGO	argonaute proteins
RISC	RNA-induced silencing complex
UTR	untranslated region
3′	3-prime
5′	5-prime
Indel	insertion/deletion
PDGFRA	platelet-derived growth factor receptor alpha
TDMD	target-directed miRNA degradation

Author details

Miruna Giurgiu, Robert Kaltenbach, Franziska Ahrend, Summer Weeks, Holly Clifton, Martin Bouldo, Vitaly Voloshin, Jiling Zhong, Siegfried Harden and Alexander Kofman*
Troy University, Troy, U.S.A.

*Address all correspondence to: akofman@troy.edu

References

[1] Arteaga-Vazquez M, Caballero-Perez J, Vielle-Calzada JP. A family of microRNAs present in plants and animals. Plant Cell. 2006;**18**(12):3355-3369

[2] Ambros V. The functions of animal microRNAs. Nature. 2004;**431**(7006):350-355

[3] Singh V. Intracellular metabolic reprogramming mediated by micro-RNAs in differentiating and proliferating cells under non-diseased conditions. Molecular Biology Reports. 2021;**48**(12):8123-8140

[4] Galagali H, Kim JK. The multifaceted roles of microRNAs in differentiation. Current Opinion in Cell Biology. 2020;**67**:118-140

[5] Zare A et al. Epigenetic modification factors and microRNAs network associated with differentiation of embryonic stem cells and induced pluripotent stem cells toward cardiomyocytes: A Review. Life (Basel). 2023;**13**(2):569. DOI: 10.3390/life13020569

[6] Souza OF, Popi AF. Role of microRNAs in B-cell compartment: Development, proliferation and Hematological diseases. Biomedicines. 2022;**10**(8):2004. DOI: 10.3390/biomedicines10082004

[7] Katsaraki K et al. MicroRNAs: Tiny regulators of gene expression with pivotal roles in Normal B-cell development and B-cell chronic lymphocytic Leukemia. Cancers (Basel). 2021;**13**(4):593. DOI: 10.3390/cancers13040593

[8] Pascale E et al. MicroRNA roles in cell reprogramming mechanisms. Cells. 2022;**11**(6):940. DOI: 10.3390/cells11060940

[9] Lettieri-Barbato D et al. MicroRNAs, long non-coding RNAs, and circular RNAs in the redox control of cell senescence. Antioxidants (Basel). 2022;**11**(3):480. DOI: 10.3390/antiox11030480

[10] Ferris WF. The role and interactions of programmed cell death 4 and its regulation by microRNA in transformed cells of the gastrointestinal tract. Frontiers in Oncology. 2022;**12**:903374

[11] Lamberti MJ et al. Damage-associated molecular patterns modulation by microRNA: Relevance on immunogenic cell death and cancer treatment outcome. Cancers (Basel). 2021;**13**(11):2566. DOI: 10.3390/cancers13112566

[12] Palumbo S et al. Emerging roles of microRNA in modulating cell-death processes in malignant glioma. Journal of Cellular Physiology. 2014;**229**(3):277-286

[13] Chen CH, Guo M, Hay BA. Identifying microRNA regulators of cell death in drosophila. Methods in Molecular Biology. 2006;**342**:229-240

[14] Bai X et al. Role of microRNA-34b-5p in cancer and injury: How does it work? Cancer Cell International. 2022;**22**(1):381

[15] Azam INA et al. Roles of microRNAs in regulating apoptosis in the pathogenesis of endometriosis. Life (Basel). 2022;**12**(9):1321. DOI: 10.3390/life12091321

[16] Matarrese P et al. The sex-related interplay between TME and cancer: On the critical role of Estrogen, MicroRNAs and autophagy. Cancers (Basel). 2021;**13**(13):3287. DOI: 10.3390/cancers13133287

[17] DeVeale B, Swindlehurst-Chan J, Blelloch R. The roles of microRNAs in mouse development. Nature Reviews Genetics. 2021;**22**(5):307-323

[18] Raza SHA et al. The role of MicroRNAs in muscle tissue development in beef cattle. Genes (Basel). 2020;**11**(3):295. DOI: 10.3390/genes11030295

[19] Boshtam M et al. Crosstalk of transcriptional regulators of adaptive immune system and microRNAs: An insight into differentiation and development. Cells. 2023;**12**(4):635. DOI: 10.3390/cells12040635

[20] Zolboot N et al. MicroRNAs instruct and maintain cell type diversity in the nervous system. Frontiers in Molecular Neuroscience. 2021;**14**:646072

[21] Mahtal N et al. MicroRNAs in kidney injury and disease. Nature Reviews Nephrology. 2022;**18**(10):643-662

[22] Rodrigues SC, Cardoso RMS, Duarte FV. Mitochondrial microRNAs: A putative role in tissue regeneration. Biology (Basel). 2021;**9**(12):486. DOI: 10.3390/biology9120486

[23] Cione E et al. Liver damage and microRNAs: An update. Current Issues in Molecular Biology. 2022;**45**(1):78-91

[24] Maries L et al. MicroRNAs-the heart of post-myocardial infarction Remodeling. Diagnostics (Basel). 2021;**11**(9):1675. DOI: 10.3390/diagnostics11091675

[25] Theofilis P et al. The impact of proangiogenic microRNA modulation on blood flow recovery following hind limb ischemia. A systematic review and meta-analysis of animal studies. Vascular Pharmacology. 2021;**141**:106906

[26] Tanase DM et al. Current knowledge of MicroRNAs (miRNAs) in acute coronary syndrome (ACS): ST-elevation myocardial infarction (STEMI). Life (Basel). 2021;**11**(10):1057. DOI: 10.3390/life11101057

[27] Vaghf A et al. The role of microRNAs in diseases and related signaling pathways. Molecular Biology Reports. 2022;**49**(7):6789-6801

[28] Humphries BA, Wang Z, Yang C. MicroRNA regulation of the small rho GTPase regulators-complexities and opportunities in targeting cancer metastasis. Cancers (Basel). 2020;**12**(5):1092. DOI: 10.3390/cancers12051092

[29] Tarlton JMR, Patterson S, Graham A. MicroRNA sequences modulated by Beta cell lipid metabolism: Implications for type 2 diabetes mellitus. Biology (Basel). 2021;**10**(6):534. DOI: 10.3390/biology10060534

[30] Wallace SR, Pagano PJ, Kracun D. MicroRNAs in the regulation of NADPH oxidases in vascular diabetic and ischemic pathologies: A case for alternate inhibitory strategies? Antioxidants (Basel). 2023;**12**(1):70. DOI: 10.3390/antiox12010070

[31] Kolodziej F et al. MicroRNAs as the sentinels of redox and hypertrophic signalling. International Journal of Molecular Sciences. 2022;**23**(23):14716. DOI: 10.3390/ijms232314716

[32] Gibson MS, Noronha-Estima C, Gama-Carvalho M. Therapeutic metabolic reprograming using microRNAs: From cancer to HIV infection. Genes (Basel). 2022;**13**(2):273. DOI: 10.3390/genes13020273

[33] Virga F et al. MicroRNA-mediated metabolic shaping of the

tumor microenvironment. Cancers (Basel). 2021;**13**(1):127. DOI: 10.3390/cancers13010127

[34] Vezza T et al. MicroRNAs and oxidative stress: An intriguing crosstalk to Be exploited in the Management of Type 2 diabetes. Antioxidants (Basel). 2021;**10**(5):802. DOI: 10.3390/antiox10050802

[35] Shiah SG, Chou ST, Chang JY. MicroRNAs: Their role in metabolism, tumor microenvironment, and therapeutic implications in head and neck squamous cell carcinoma. Cancers (Basel). 2021;**13**(22):5604. DOI: 10.3390/cancers13225604

[36] Azizi M, Othman I, Naidu R. The role of MicroRNAs in lung cancer metabolism. Cancers (Basel). 2021;**13**(7):1716. DOI: 10.3390/cancers13071716

[37] Wai Hon K et al. Insights into the role of microRNAs in colorectal cancer (CRC). Metabolism. Cancers (Basel). 2020;**12**(9):2462. DOI: 10.3390/cancers12092462

[38] Ortega MA et al. The regulatory role of mitochondrial MicroRNAs (MitomiRs) in breast cancer: Translational implications present and future. Cancers (Basel). 2020;**12**(9):2443. DOI: 10.3390/cancers12092443

[39] Tan WL et al. An insight into the associations between microRNA expression and mitochondrial functions in cancer cell and cancer stem cell. Molecular Biology Reports. 2023;**50**(6):5395-5405

[40] Macias S, Michlewski G, Cáceres JF. Hormonal regulation of MicroRNA biogenesis. Molecular Cell. 2009;**36**(2):172-173

[41] Kim IK et al. The role of Epicardial adipose tissue-derived MicroRNAs in the regulation of cardiovascular disease: A narrative review. Biology (Basel). 2023;**12**(4):498. DOI: 10.3390/biology12040498

[42] Brennan GP, Henshall DC. MicroRNAs as regulators of brain function and targets for treatment of epilepsy. Nature Reviews Neurology. 2020;**16**(9):506-519

[43] Cao DD, Li L, Chan WY. MicroRNAs: Key regulators in the central nervous system and their implication in neurological diseases. International Journal of Molecular Sciences. 2016;**17**(6):842. DOI: 10.3390/ijms17060842

[44] Brites D. Regulatory function of microRNAs in microglia. Glia. 2020;**68**(8):1631-1642

[45] Lee A, Kim SN. Serum MicroRNA on inflammation: A literature review of mouse model studies. Biomarkers. 2020;**25**(7):513-524

[46] Maranini B et al. microRNAs and inflammatory immune response in SARS-CoV-2 infection: A narrative review. Life (Basel). 2022;**12**(2):288. DOI: 10.3390/life12020288

[47] Bi K et al. MicroRNAs regulate intestinal immunity and gut microbiota for gastrointestinal health: A comprehensive review. Genes (Basel). 2020;**11**(9):1075. DOI: 10.3390/genes11091075

[48] Allegra A et al. Interactions between the MicroRNAs and microbiota in cancer development: Roles and therapeutic opportunities. Cancers (Basel). 2020;**12**(4):805. DOI: 10.3390/cancers12040805

[49] Zhou H et al. MicroRNAs with multiple targets of immune checkpoints, as a potential sensitizer for immune checkpoint inhibitors

in breast cancer treatment. Cancers (Basel). 2023;**15**(3):824. DOI: 10.3390/cancers15030824

[50] Mourenza A et al. Understanding microRNAs in the context of infection to find new treatments against human bacterial pathogens. Antibiotics (Basel). 2022;**11**(3):356. DOI: 10.3390/antibiotics11030356

[51] Sarshar M et al. Fecal microRNAs as innovative biomarkers of intestinal diseases and effective players in host-microbiome interactions. Cancers (Basel). 2020;**12**(8):2174. DOI: 10.3390/cancers12082174

[52] Cho S, Tai JW, Lu LF. MicroRNAs and their Targetomes in tumor-immune communication. Cancers (Basel). 2020;**12**(8):2025. DOI: 10.3390/cancers12082025

[53] De Silva S, Tennekoon KH, Karunanayake EH. Interaction of gut microbiome and host microRNAs with the occurrence of colorectal and breast cancer and their impact on patient immunity. Oncotargets and Therapy. 2021;**14**:5115-5129

[54] Kofman AV et al. The p53-microRNA-34a axis regulates cellular entry receptors for tumor-associated human herpes viruses. Medical Hypotheses. 2013;**81**(1):62-67

[55] Cullen BR. Viral and cellular messenger RNA targets of viral microRNAs. Nature. 2009;**457**(7228):421-425

[56] Zhang L, Yu J, Liu Z. MicroRNAs expressed by human cytomegalovirus. Virology Journal. 2020;**17**(1):34

[57] Zhan S, Wang Y, Chen X. RNA virus-encoded microRNAs: Biogenesis, functions and perspectives on application. ExRNA. 2020;**2**(1):15

[58] Abdalla AE et al. Human cytomegalovirus-encoded MicroRNAs: A master regulator of latent infection. Infection, Genetics and Evolution. 2020;**78**:104119

[59] Kandeel M. Oncogenic viruses-encoded microRNAs and their role in the progression of cancer: Emerging targets for antiviral and anticancer therapies. Pharmaceuticals (Basel). 2023;**16**(4):485. DOI: 10.3390/ph16040485

[60] Yu M et al. microRNA, a subtle indicator of human cytomegalovirus against host immune cells. Vaccines (Basel). 2022;**10**(2):144. DOI: 10.3390/vaccines10020144

[61] Panigrahi M, Palmer MA, Wilson JA. MicroRNA-122 regulation of HCV infections: Insights from studies of miR-122-independent replication. Pathogens. 2022;**11**(9):1005. DOI: 10.3390/pathogens11091005

[62] Li Y et al. The role of MicroRNA in DNA damage response. Frontiers in Genetics. 2022;**13**:850038

[63] Visser H, Thomas AD. MicroRNAs and the DNA damage response: How is cell fate determined? DNA Repair (Amst). 2021;**108**:103245

[64] Orafidiya F et al. Crosstalk between long non coding RNAs, microRNAs and DNA damage repair in prostate cancer: New therapeutic opportunities? Cancers (Basel). 2022;**14**(3):755. DOI: 10.3390/cancers14030755

[65] Gajek A et al. Current implications of microRNAs in genome stability and stress responses of ovarian cancer. Cancers (Basel). 2021;**13**(11):2690. DOI: 10.3390/cancers13112690

[66] Szatkowska M, Krupa R. Regulation of DNA damage response and

homologous recombination repair by microRNA in human cells exposed to ionizing radiation. Cancers (Basel). 2020;**12**(7):1838. DOI: 10.3390/cancers12071838

[67] Pellegrini L et al. MicroRNAs in cancer treatment-induced cardiotoxicity. Cancers (Basel). 2020;**12**(3):704. DOI: 10.3390/cancers12030704

[68] Shi X, Yang H, Birchler JA. MicroRNAs play regulatory roles in genomic balance. BioEssays. 2023;**45**(2):e2200187

[69] Huppi K et al. Genomic instability and mouse microRNAs. Toxicology Mechanisms and Methods. 2011;**21**(4):325-333

[70] Devara D, Choudhary Y, Kumar S. Role of MicroRNA-502-3p in human diseases. Pharmaceuticals (Basel). 2023;**16**(4):532. DOI: 10.3390/ph16040532

[71] Zhao Y et al. New insights into the functions of MicroRNAs in cardiac fibrosis: From mechanisms to therapeutic strategies. Genes (Basel). 2022;**13**(8):1390. DOI: 10.3390/genes13081390

[72] Stunf, Pukl S. MicroRNA of epithelial to mesenchymal transition in Fuchs' endothelial corneal dystrophy. Genes (Basel). 2022;**13**(10):1711. DOI: 10.3390/genes13101711

[73] Ruiz-Manriquez LM et al. A brief review on the regulatory roles of MicroRNAs in cystic diseases and their use as potential biomarkers. Genes (Basel). 2022;**13**(2):191. DOI: 10.3390/genes13020191

[74] Ramirez AE et al. MicroRNA: A linking between astrocyte dysfunction, mild cognitive impairment, and neurodegenerative diseases. Life (Basel). 2022;**12**(9):1439. DOI: 10.3390/life12091439

[75] Moriondo G et al. Obstructive sleep Apnea: A look towards micro-RNAs as biomarkers of the future. Biology (Basel). 2023;**12**(1):66. DOI: 10.3390/biology12010066

[76] Lee C, Han J, Jung Y. Pathological contribution of extracellular vesicles and their MicroRNAs to progression of chronic liver disease. Biology (Basel). 2022;**11**(5):637. DOI: 10.3390/biology11050637

[77] Iulian, Stanciugelu S et al. Osteoarthritis and microRNAs: Do they provide novel insights into the pathophysiology of this degenerative disorder? Life (Basel). 2022;**12**(11):1914. DOI: 10.3390/life12111914

[78] Homorogan C et al. Uncovering the roles of MicroRNAs in major depressive disorder: From candidate diagnostic biomarkers to treatment response indicators. Life (Basel). 2021;**11**(10):1073. DOI: 10.3390/life11101073

[79] De Benedittis G et al. Emerging role of microRNAs and long non-coding RNAs in Sjogren's syndrome. Genes (Basel). 2021;**12**(6):903. DOI: 10.3390/genes12060903

[80] Chiti E et al. MicroRNAs in hypertrophic, Arrhythmogenic and dilated cardiomyopathy. Diagnostics (Basel). 2021;**11**(9):1720. DOI: 10.3390/diagnostics11091720

[81] Barbiera A et al. Nutrition and microRNAs: Novel insights to fight sarcopenia. Antioxidants (Basel). 2020;**9**(10):951. DOI: 10.3390/antiox9100951

[82] Martins B et al. Extracellular vesicles and MicroRNA: Putative role in diagnosis

and treatment of diabetic retinopathy. Antioxidants (Basel). 2020;**9**(8):705. DOI: 10.3390/antiox9080705

[83] De Palma FDE et al. The multifaceted roles of MicroRNAs in cystic fibrosis. Diagnostics (Basel). 2020;**10**(12):1102. DOI: 10.3390/diagnostics10121102

[84] Suzuki HI. Roles of MicroRNAs in disease biology. JMA Journal. 2023;**6**(2):104-113

[85] Kaller M et al. Analysis of the p53/microRNA network in cancer. Advances in Experimental Medicine and Biology. 2022;**1385**:187-228

[86] Dayakar A, Shanmukha KD, Kalangi SK. Spectrum of microRNAs and their target genes in cancer: Intervention in diagnosis and therapy. Molecular Biology Reports. 2022;**49**(7):6827-6846

[87] Semina EV et al. MicroRNAs in cancer: From gene expression regulation to the metastatic niche reprogramming. Biochemistry (Mosc). 2021;**86**(7):785-799

[88] Angius A et al. Modulatory role of microRNAs in triple negative breast cancer with basal-like phenotype. Cancers (Basel). 2020;**12**(11):3298. DOI: 10.3390/cancers12113298

[89] Alshamrani AA. Roles of microRNAs in ovarian cancer tumorigenesis: Two decades later, what have we learned? Frontiers in Oncology. 2020;**10**:1084

[90] Sell MC et al. MicroRNAs in cancer metastasis: Biological and therapeutic implications. Expert Reviews in Molecular Medicine. 2023;**25**:e14

[91] Pekarek L et al. An overview of the role of MicroRNAs on carcinogenesis: A focus on cell cycle, angiogenesis and metastasis. International Journal of Molecular Sciences. 2023;**24**(8):7268. DOI: 10.3390/ijms24087268

[92] Pan W et al. p53/MicroRNA-34 axis in cancer and beyond. Heliyon. 2023;**9**(4):e15155

[93] Sevcikova A et al. Clinical significance of microRNAs in hematologic malignancies and hematopoietic stem cell transplantation. Cancers (Basel). 2023;**15**(9):2658. DOI: 10.3390/cancers15092658

[94] Tran F et al. MicroRNA-gene interactions impacted by toxic metal(oid)s during EMT and carcinogenesis. Cancers (Basel). 2022;**14**(23):5818. DOI: 10.3390/cancers14235818

[95] Sobolewski C, Dubuquoy L, Legrand N. MicroRNAs, Tristetraprolin family members and HuR: A complex interplay controlling cancer-related processes. Cancers (Basel). 2022;**14**(14):3516. DOI: 10.3390/cancers14143516

[96] Setlai BP et al. MicroRNA interrelated epithelial mesenchymal transition (EMT) in glioblastoma. Genes (Basel). 2022;**13**(2):244. DOI: 10.3390/genes13020244

[97] Rajtmajerova M et al. Long non-coding RNA and microRNA interplay in colorectal cancer and their effect on the tumor microenvironment. Cancers (Basel). 2022;**14**(21):5450. DOI: 10.3390/cancers14215450

[98] Prinz C et al. Emerging role of microRNA dysregulation in diagnosis and prognosis of extrahepatic cholangiocarcinoma. Genes (Basel). 2022;**13**(8):1479. DOI: 10.3390/genes13081479

[99] Pecorelli A, Valacchi G. Oxidative-stress-sensitive microRNAs in

UV-promoted development of melanoma. Cancers (Basel). 2022;**14**(13):3224. DOI: 10.3390/cancers14133224

[100] Koustas E et al. The emerging role of MicroRNAs and autophagy mechanism in pancreatic cancer progression: Future therapeutic approaches. Genes (Basel). 2022;**13**(10):1868. DOI: 10.3390/genes13101868

[101] Wang X et al. Effects of CAF-derived MicroRNA on tumor biology and clinical applications. Cancers (Basel). 2021;**13**(13):3160. DOI: 10.3390/cancers13133160

[102] Tomei S et al. Cancer stem cells are possible key players in regulating anti-tumor immune responses: The role of Immunomodulating molecules and MicroRNAs. Cancers (Basel). 2021;**13**(7):1674. DOI: 10.3390/cancers13071674

[103] Morishita A et al. MicroRNAs in the pathogenesis of hepatocellular carcinoma: A review. Cancers (Basel). 2021;**13**(3):514. DOI: 10.3390/cancers13030514

[104] Fitriana M et al. Roles of microRNAs in regulating cancer Stemness in head and neck cancers. Cancers (Basel). 2021;**13**(7):1742. DOI: 10.3390/cancers13071742

[105] Kipkeeva F et al. MicroRNA in gastric cancer development: Mechanisms and biomarkers. Diagnostics (Basel). 2020;**10**(11):891. DOI: 10.3390/diagnostics10110891

[106] Gluud M et al. MicroRNAs in the pathogenesis, diagnosis, prognosis and targeted treatment of cutaneous T-cell lymphomas. Cancers (Basel). 2020;**12**(5):1229. DOI: 10.3390/cancers12051229

[107] Soond SM et al. Integrative p53, micro-RNA and Cathepsin protease Co-regulatory expression networks in cancer. Cancers (Basel). 2020;**12**(11):3454. DOI: 10.3390/cancers12113454

[108] Aali M et al. Evaluating the role of microRNAs alterations in oral squamous cell carcinoma. Gene. 2020;**757**:144936

[109] Wang N et al. MicroRNA-149: A review of its role in digestive system cancers. Pathology, Research and Practice. 2020;**216**(12):153266

[110] Zietara KJ et al. The importance of selected dysregulated microRNAs in diagnosis and prognosis of childhood B-cell precursor acute lymphoblastic Leukemia. Cancers (Basel). 2023;**15**(2):428. DOI: 10.3390/cancers15020428

[111] Teo AYT, Lim VY, Yang VS. MicroRNAs in the pathogenesis, prognostication and prediction of treatment resistance in soft tissue sarcomas. Cancers (Basel). 2020;**12**:1229. DOI: 10.3390/cancers12051229

[112] Psilopatis I et al. The role of MicroRNAs in uterine Leiomyosarcoma diagnosis and treatment. Cancers (Basel). 2023;**15**(9):2420. DOI: 10.3390/cancers15092420

[113] Maldonado E et al. Role of the mediator complex and MicroRNAs in breast cancer Etiology. Genes (Basel). 2022;**13**(2):234. DOI: 10.3390/genes13020234

[114] Fogazzi V et al. The role of MicroRNAs in HER2-positive breast cancer: Where we are and future prospective. Cancers (Basel). 2022;**14**(21):5326. DOI: 10.3390/cancers14215326

[115] Bocchetti M et al. MicroRNAs' crucial role in salivary gland Cancers' onset and prognosis. Cancers (Basel).

2022;**14**(21):5304. DOI: 10.3390/cancers14215304

[116] Notarte KI et al. MicroRNA and other non-coding RNAs in Epstein-Barr virus-associated cancers. Cancers (Basel). 2021;**13**(15):3909. DOI: 10.3390/cancers13153909

[117] Bevacqua E et al. Role of MicroRNAs in the development and progression of the four Medulloblastoma subgroups. Cancers (Basel). 2021;**13**(24):6323. DOI: 10.3390/cancers13246323

[118] Sidorkiewicz I et al. Insulin resistance and endometrial cancer: Emerging role for microRNA. Cancers (Basel). 2020;**12**(9):2559. DOI: 10.3390/cancers12092559

[119] Hitu L et al. MicroRNA in papillary thyroid carcinoma: A systematic review from 2018 to June 2020. Cancers (Basel). 2020;**12**(11):3118. DOI: 10.3390/cancers12113118

[120] Khan P et al. MicroRNA-1: Diverse role of a small player in multiple cancers. Seminars in Cell & Developmental Biology. 2022;**124**:114-126

[121] Grzywa TM, Klicka K, Wlodarski PK. Regulators at every step-how microRNAs drive tumor cell invasiveness and metastasis. Cancers (Basel). 2020;**12**(12):3709. DOI: 10.3390/cancers12123709

[122] Siegl F et al. The significance of MicroRNAs in the molecular pathology of brain metastases. Cancers (Basel). 2022;**14**(14):3386. DOI: 10.3390/cancers14143386

[123] Rodriguez, Calleja L et al. The p53 family members p63 and p73 roles in the metastatic dissemination: Interactions with microRNAs and TGFbeta pathway. Cancers (Basel). 2022;**14**(23):5948. DOI: 10.3390/cancers14235948

[124] Haider MT, Smit DJ, Taipaleenmaki H. MicroRNAs: Emerging regulators of metastatic bone disease in breast cancer. Cancers (Basel). 2022;**14**(3):729. DOI: 10.3390/cancers14030729

[125] Tang LB et al. Exosomal microRNAs: Pleiotropic impacts on breast cancer metastasis and their clinical perspectives. Biology (Basel). 2021;**10**(4):307. DOI: 10.3390/biology10040307

[126] Oh-Hohenhorst SJ, Lange T. Role of metastasis-related microRNAs in prostate cancer progression and treatment. Cancers (Basel). 2021;**13**(17):4492. DOI: 10.3390/cancers13174492

[127] Hammouz RY et al. MicroRNAs: Their role in metastasis, angiogenesis, and the potential for biomarker utility in bladder carcinomas. Cancers (Basel). 2021;**13**(4):891. DOI: 10.3390/cancers13040891

[128] Wood SL, Brown JE. Personal medicine and bone metastases: Biomarkers, micro-RNAs and bone metastases. Cancers (Basel). 2020;**12**(8):2109. DOI: 10.3390/cancers12082109

[129] Weidle UH et al. MicroRNAs involved in metastasis of hepatocellular carcinoma: Target candidates, functionality and efficacy in animal models and prognostic relevance. Cancer Genomics Proteomics. 2020;**17**(1):1-21

[130] Santos JMO et al. The emerging role of MicroRNAs and other non-coding RNAs in cancer cachexia. Cancers (Basel). 2020;**12**(4):1004. DOI: 10.3390/cancers12041004

[131] Pavlikova L et al. The roles of microRNAs in cancer multidrug resistance.

Cancers (Basel). 2022;**14**(4):1090. DOI: 10.3390/cancers14041090

[132] Mahinfar P et al. The role of microRNAs in multidrug resistance of glioblastoma. Cancers (Basel). 2022;**14**(13):3217. DOI: 10.3390/cancers14133217

[133] Cuttano R, Afanga MK, Bianchi F. MicroRNAs and drug resistance in non-small cell lung cancer: Where are we now and where are we going. Cancers (Basel). 2022;**14**(23):5731. DOI: 10.3390/cancers14235731

[134] Crudele F et al. The molecular networks of microRNAs and their targets in the drug resistance of colon carcinoma. Cancers (Basel). 2021;**13**(17):4355. DOI: 10.3390/cancers13174355

[135] Cosentino G et al. Breast cancer drug resistance: Overcoming the challenge by capitalizing on MicroRNA and tumor microenvironment interplay. Cancers (Basel). 2021;**13**(15)

[136] Lampis A et al. MicroRNAs as mediators of drug resistance mechanisms. Current Opinion in Pharmacology. 2020;**54**:44-50

[137] Funamizu N et al. microRNAs associated with gemcitabine resistance via EMT, TME, and drug metabolism in pancreatic cancer. Cancers (Basel). 2023;**15**(4)

[138] Masadah R et al. The role of microRNAs in the cisplatin- and radio-resistance of cervical cancer. Cancers (Basel). 2021;**13**(5)

[139] Podralska M et al. Non-coding RNAs in cancer Radiosensitivity: MicroRNAs and lncRNAs as regulators of radiation-induced Signaling pathways. Cancers (Basel). 2020;**12**(6)

[140] Bozzato AM et al. MicroRNAs related to TACE treatment response: A review of the literature from a radiological point of view. Diagnostics (Basel). 2022;**12**(2)

[141] Sukocheva OA et al. Perspectives of using microRNA-loaded nanocarriers for epigenetic reprogramming of drug resistant colorectal cancers. Seminars in Cancer Biology. 2022;**86**(Pt 2):358-375

[142] Arghiani N, Shah K. Modulating microRNAs in cancer: Next-generation therapies. Cancer Biology & Medicine. 2021;**19**(3):289-304

[143] Karkhane M et al. Oncogenesis and tumor inhibition by MicroRNAs and its potential therapeutic applications: A systematic review. MicroRNA. 2020;**9**(3):198-215

[144] Chen L et al. Premature MicroRNA-based therapeutic: A "one-two punch" against cancers. Cancers (Basel). 2020;**12**(12)

[145] Abd-Aziz N, Kamaruzman NI, Poh CL. Development of MicroRNAs as potential therapeutics against cancer. Journal of Oncology. 2020;**2020**:8029721

[146] Samad AFA, Kamaroddin MF. Innovative approaches in transforming microRNAs into therapeutic tools. Wiley Interdisciplinary Reviews. RNA. 2023;**14**(1):e1768

[147] Liu J et al. SALL4 and microRNA: The role of Let-7. Genes (Basel). 2021;**12**(9)

[148] Li WJ et al. MicroRNA-34a: Potent tumor suppressor, cancer stem cell inhibitor, and potential anticancer therapeutic. Frontiers in Cell and Development Biology. 2021;**9**:640587

[149] Li WJ et al. Prostate cancer stem cells, and therapeutic development. Cancers (Basel). 2022;**14**(18)

[150] Kara G et al. miRacle of microRNA-driven cancer Nanotherapeutics. Cancers (Basel). 2022;**14**(15)

[151] Genedy HH, Delair T, Montembault A. Chitosan Based MicroRNA Nanocarriers. Pharmaceuticals (Basel). 2022;**15**(9)

[152] Romano G, Acunzo M, Nana-Sinkam P. microRNAs as Novel Therapeutics in Cancer. Cancers (Basel). 2021;**13**(7)

[153] Reda, El Sayed S et al. MicroRNA therapeutics in cancer: Current advances and challenges. Cancers (Basel). 2021;**13**(11)

[154] Shadbad MA et al. A systematic review on the therapeutic potentiality of PD-L1-inhibiting MicroRNAs for triple-negative breast cancer: Toward single-cell sequencing-guided biomimetic delivery. Genes (Basel). 2021;**12**(8)

[155] Ohtsuka M et al. Circulating MicroRNAs in Gastrointestinal Cancer. Cancers (Basel). 2021;**13**:13

[156] Niccolini B et al. Opportunities offered by graphene nanoparticles for MicroRNAs delivery for amyotrophic lateral sclerosis treatment. Materials (Basel). 2021;**15**(1)

[157] Jung E et al. MicroRNA-based therapeutics for drug-resistant colorectal cancer. Pharmaceuticals (Basel). 2021;**14**(2)

[158] Wong GL, Abu Jalboush S, Lo HW. Exosomal MicroRNAs and organotropism in breast cancer metastasis. Cancers (Basel). 2020;**12**(7)

[159] Balachandran AA et al. Therapeutically significant MicroRNAs in primary and metastatic brain malignancies. Cancers (Basel). 2020;**12**(9)

[160] Imedio L et al. MicroRNAs in rectal cancer: Functional significance and promising therapeutic value. Cancers (Basel). 2020;**12**(8)

[161] Hsieh PL et al. MicroRNAs as Theranostics targets in Oral carcinoma stem cells. Cancers (Basel). 2020;**12**(2)

[162] Fortunato O, Iorio MV. The therapeutic potential of MicroRNAs in cancer: Illusion or opportunity? Pharmaceuticals (Basel). 2020;**13**(12)

[163] Forterre A et al. A comprehensive review of cancer MicroRNA therapeutic delivery strategies. Cancers (Basel). 2020;**12**(7)

[164] Eckburg A et al. Oligonucleotides and microRNAs targeting telomerase subunits in cancer therapy. Cancers (Basel). 2020;**12**(9)

[165] Liu D et al. Drugging the "undruggable" microRNAs. Cellular and Molecular Life Sciences. 2021;**78**(5):1861-1871

[166] Chaudhry T, Coxon CR, Ross K. Trading places: Peptide and small molecule alternatives to oligonucleotide-based modulation of microRNA expression. Drug Discovery Today. 2022;**27**(11):103337

[167] Van Meter EN, Onyango JA, Teske KA. A review of currently identified small molecule modulators of microRNA function. European Journal of Medicinal Chemistry. 2020;**188**:112008

[168] Ma L et al. MicroRNAs as mediators of adipose thermogenesis and potential therapeutic targets for obesity. Biology (Basel). 2022;**11**(11)

[169] Weidle UH et al. Gastric cancer: Identification of microRNAs inhibiting Druggable targets and mediating

efficacy in preclinical In vivo models. Cancer Genomics Proteomics. 2021;**18**(4):497-514

[170] Weidle UH et al. Down-regulated MicroRNAs in gastric carcinoma may Be targets for therapeutic intervention and replacement therapy. Anticancer Research. 2021;**41**(9):4185-4202

[171] Weidle UH, Birzele F. Bladder cancer-related microRNAs with in vivo efficacy in preclinical models. Cancer Diagnosis & Prognosis. 2021;**1**(4):245-263

[172] Doldi V, El Bezawy R, Zaffaroni N. MicroRNAs as epigenetic determinants of treatment response and potential therapeutic targets in prostate cancer. Cancers (Basel). 2021;**13**(10)

[173] Cannataro R et al. Polyphenols in the Mediterranean diet: From dietary sources to microRNA modulation. Antioxidants (Basel). 2021;**10**(2)

[174] Chehade M et al. Key MicroRNA's and their Targetome in adrenocortical cancer. Cancers (Basel). 2020;**12**(8)

[175] Christou N et al. Circulating tumour cells, circulating tumour DNA and circulating tumour miRNA in blood assays in the different steps of colorectal cancer management, a review of the evidence in 2019. BioMed Research International. 2019;**2019**:5953036

[176] Olivieri F et al. Circulating miRNAs and miRNA shuttles as biomarkers: Perspective trajectories of healthy and unhealthy aging. Mechanisms of Ageing and Development. 2017;**165**(Pt B):162-170

[177] Rocchi A et al. MicroRNAs: An update of applications in forensic science. Diagnostics (Basel). 2020;**11**(1)

[178] Regouc M et al. Non-coding microRNAs as novel potential tumor markers in testicular cancer. Cancers (Basel). 2020;**12**(3)

[179] Glynn CL. Potential applications of microRNA profiling to forensic investigations. RNA. 2020;**26**(1):1-9

[180] Ghosh RD, Pattatheyil A, Roychoudhury S. Functional landscape of dysregulated MicroRNAs in Oral squamous cell carcinoma: Clinical implications. Frontiers in Oncology. 2020;**10**:619

[181] Varela N et al. The current state of MicroRNAs as restenosis biomarkers. Frontiers in Genetics. 2019;**10**:1247

[182] Ahadi A. A systematic review of microRNAs as potential biomarkers for diagnosis and prognosis of gastric cancer. Immunogenetics. 2021;**73**(2):155-161

[183] Yang IP et al. MicroRNAs as predictive biomarkers in patients with colorectal cancer receiving chemotherapy or chemoradiotherapy: A narrative literature review. Cancers (Basel). 2023;**15**(5)

[184] Salamanna F et al. Sharing circulating micro-RNAs between osteoporosis and sarcopenia: A systematic review. Life (Basel). 2023;**13**(3)

[185] Fawzy MS et al. MicroRNA-155 and disease-related Immunohistochemical parameters in cutaneous melanoma. Diagnostics (Basel). 2023;**13**(6)

[186] Zhang J et al. Recent advances in the roles of MicroRNA and MicroRNA-based diagnosis in neurodegenerative diseases. Biosensors (Basel). 2022;**12**(12)

[187] Wielogorska M et al. MicroRNA profile alterations in parathyroid carcinoma: Latest updates and perspectives. Cancers (Basel). 2022;**14**(4)

[188] Vykoukal J et al. Contributions of circulating microRNAs for early detection of lung cancer. Cancers (Basel). 2022;**14**(17)

[189] Shaw P et al. A clinical investigation on the Theragnostic effect of MicroRNA biomarkers for survival outcome in cervical cancer: A PRISMA-P compliant protocol for systematic review and comprehensive meta-analysis. Genes (Basel). 2022;**13**(3)

[190] Le TAH, Lao TD. Circulating microRNAs as the potential diagnostic and prognostic biomarkers for nasopharyngeal carcinoma. Genes (Basel). 2022;**13**(7)

[191] Kyriakidis I, Kyriakidis K, Tsezou A. MicroRNAs and the diagnosis of childhood acute lymphoblastic Leukemia: Systematic review, meta-analysis and Re-analysis with novel small RNA-Seq tools. Cancers (Basel). 2022;**14**(16)

[192] Juracek J et al. Urinary microRNAs and their significance in prostate cancer diagnosis: A 5-year update. Cancers (Basel). 2022;**14**(13)

[193] Davies M, Davey MG, Miller N. The potential of MicroRNAs as clinical biomarkers to aid ovarian cancer diagnosis and treatment. Genes (Basel). 2022;**13**(11)

[194] Dabi Y et al. Clues for improving the pathophysiology knowledge for endometriosis using serum micro-RNA expression. Diagnostics (Basel). 2022;**12**(1)

[195] Bevacqua E et al. The potential of MicroRNAs as non-invasive prostate cancer biomarkers: A systematic literature review based on a machine learning approach. Cancers (Basel). 2022;**14**(21)

[196] Aghiorghiesei O et al. The world of Oral cancer and its risk factors viewed from the aspect of MicroRNA expression patterns. Genes (Basel). 2022;**13**(4)

[197] Zhao Z et al. Fecal microRNAs, Fecal microRNA panels, or combinations of Fecal microRNAs with Fecal Hemoglobin for early detection of colorectal cancer and its precursors: A systematic review. Cancers (Basel). 2021;**14**(1)

[198] Weidle UH, Birzele F, Nopora A. microRNAs promoting growth of gastric cancer xenografts and correlation to clinical prognosis. Cancer Genomics Proteomics. 2021;**18**(1):1-15

[199] Turai PI et al. MicroRNAs, long non-coding RNAs, and circular RNAs: Potential biomarkers and therapeutic targets in Pheochromocytoma/Paraganglioma. Cancers (Basel). 2021;**13**(7)

[200] Tito C et al. Circulating microRNAs from the molecular mechanisms to clinical biomarkers: A focus on the clear cell renal cell carcinoma. Genes (Basel). 2021;**12**(8)

[201] Shaw P et al. A clinical update on the prognostic effect of microRNA biomarkers for survival outcome in nasopharyngeal carcinoma: A systematic review and meta-analysis. Cancers (Basel). 2021;**13**(17)

[202] Sequeira JP et al. Unveiling the world of circulating and Exosomal microRNAs in renal cell carcinoma. Cancers (Basel). 2021;**13**(21)

[203] Rzepka-Migut B, Paprocka J. Prospects and limitations related to the use of MicroRNA as a biomarker of epilepsy in children: A systematic review. Life (Basel). 2021;**11**(1)

[204] Richard V et al. MicroRNAs in molecular classification and pathogenesis

of breast Tumors. Cancers (Basel). 2021;**13**(21)

[205] Prinz C, Mese K, Weber D. MicroRNA changes in gastric carcinogenesis: Differential dysregulation during helicobacter pylori and EBV infection. Genes (Basel). 2021;**12**(4)

[206] Osan C et al. The connection between MicroRNAs and Oral cancer pathogenesis: Emerging biomarkers in Oral cancer management. Genes (Basel). 2021;**12**(12)

[207] MacKenzie SM et al. Circulating microRNAs as diagnostic markers in primary Aldosteronism. Cancers (Basel). 2021;**13**(21)

[208] Low SS et al. Recent Progress in nanomaterials modified electrochemical biosensors for the detection of MicroRNA. Micromachines (Basel). 2021;**12**(11)

[209] Lao TD, Le TAH. Data integration reveals the potential biomarkers of circulating MicroRNAs in osteoarthritis. Diagnostics (Basel). 2021;**11**(3)

[210] Jiang Y et al. MicroRNAs as potential biomarkers for exercise-based cancer rehabilitation in cancer survivors. Life (Basel). 2021;**11**(12)

[211] Huang PS et al. Functional and clinical significance of dysregulated microRNAs in liver cancer. Cancers (Basel). 2021;**13**(21)

[212] Gayosso-Gomez LV, Ortiz-Quintero B. Circulating MICRORNAS in blood and other body fluids as biomarkers for diagnosis, prognosis, and therapy response in lung cancer. Diagnostics (Basel). 2021:**11**(3):421. DOI: 10.3390/diagnostics11030421

[213] Favier A et al. MicroRNA as epigenetic modifiers in endometrial cancer: A systematic review. Cancers (Basel). 2021;**13**(5)

[214] Desantis V et al. MicroRNAs as a potential new preventive approach in the transition from asymptomatic to symptomatic multiple myeloma disease. Cancers (Basel). 2021;**13**(15)

[215] Cione E et al. Exosome microRNAs in metabolic syndrome as tools for the early monitoring of diabetes and possible therapeutic options. Pharmaceuticals (Basel). 2021;**14**(12)

[216] Chorley BN et al. Methodological considerations for measuring biofluid-based microRNA biomarkers. Critical Reviews in Toxicology. 2021;**51**(3):264-282

[217] Cavallari I et al. The miR-200 family of microRNAs: Fine tuners of epithelial-mesenchymal transition and circulating cancer biomarkers. Cancers (Basel). 2021;**13**(23)

[218] Borgia F et al. Involvement of microRNAs as a response to phototherapy and photodynamic therapy: A literature review. Antioxidants (Basel). 2021;**10**(8)

[219] Andrikopoulou A et al. MicroRNAs as potential predictors of response to CDK4/6 inhibitor treatment. Cancers (Basel). 2021;**13**(16)

[220] Al-Akhrass H, Christou N. The clinical assessment of MicroRNA diagnostic, prognostic, and Theranostic value in colorectal cancer. Cancers (Basel). 2021;**13**(12)

[221] El Aamri M et al. Electrochemical biosensors for detection of MicroRNA as a cancer biomarker: Pros and cons. Biosensors (Basel). 2020;**10**(11)

[222] Silaghi CA et al. The prognostic value of MicroRNAs in thyroid cancers-a

systematic review and meta-analysis. Cancers (Basel). 2020;**12**(9)

[223] Izzotti A et al. Predicting response to neoadjuvant therapy in colorectal cancer patients the role of messenger-and micro-RNA profiling. Cancers (Basel). 2020;**12**(6)

[224] Maiese A et al. MicroRNAs as useful tools to estimate time since death. A systematic review of current literature. Diagnostics (Basel). 2021;**11**(1)

[225] Witwer KW. XenomiRs and miRNA homeostasis in health and disease: Evidence that diet and dietary miRNAs directly and indirectly influence circulating miRNA profiles. RNA Biology. 2012;**9**(9):1147-1154

[226] Soplinska A et al. MicroRNAs as biomarkers of systemic changes in response to endurance exercise-a comprehensive review. Diagnostics (Basel). 2020;**10**(10)

[227] Dufresne S et al. A review of physical activity and circulating miRNA expression: Implications in cancer risk and progression. Cancer Epidemiology, Biomarkers & Prevention. 2018;**27**(1):11-24

[228] Lewis BP et al. Prediction of mammalian microRNA targets. Cell. 2003;**115**(7):787-798

[229] Li Z, Xu R, Li N. MicroRNAs from plants to animals, do they define a new messenger for communication? Nutrition & Metabolism (London). 2018;**15**:68

[230] Landi D, Gemignani F, Landi S. Role of variations within microRNA-binding sites in cancer. Mutagenesis. 2012;**27**(2):205-210

[231] Kasinski AL, Slack FJ. Epigenetics and genetics. MicroRNAs en route to the clinic: Progress in validating and targeting microRNAs for cancer therapy. Nature Reviews Cancer. 2011;**11**(12):849-864

[232] de Rooij LA et al. The microRNA lifecycle in health and cancer. Cancers (Basel). 2022;**14**(23)

[233] Bartel DP. MicroRNAs: Target recognition and regulatory functions. Cell. 2009;**136**(2):215-233

[234] Liu J, Liu F. The yin and Yang function of microRNAs in insulin signalling and cancer. RNA Biology. 2021;**18**(1):24-32

[235] Sindhu KJ, Venkatesan N, Karunagaran D. MicroRNA Interactome multiomics characterization for cancer research and personalized medicine: An expert review. OMICS. 2021;**25**(9):545-566

[236] Komatsu S, Kitai H, Suzuki HI. Network regulation of microRNA biogenesis and target interaction. Cell. 2023;**12**(2)

[237] Skrzypek K, Majka M. Interplay among SNAIL transcription factor, MicroRNAs, long non-coding RNAs, and circular RNAs in the regulation of tumor growth and metastasis. Cancers (Basel). 2020;**12**(1)

[238] Poli V, Secli L, Avalle L. The Microrna-143/145 cluster in tumors: A matter of where and when. Cancers (Basel). 2020;**12**(3)

[239] Perri P et al. A focus on regulatory networks linking MicroRNAs, transcription factors and target genes in neuroblastoma. Cancers (Basel). 2021;**13**(21)

[240] Chen K et al. Polymorphisms in microRNA targets: A gold mine for molecular epidemiology. Carcinogenesis. 2008;**29**(7):1306-1311

[241] Sethupathy P, Collins FS. MicroRNA target site polymorphisms and human disease. Trends in Genetics. 2008;**24**(10):489-497

[242] Pelletier C, Weidhaas JB. MicroRNA binding site polymorphisms as biomarkers of cancer risk. Expert Review of Molecular Diagnostics. 2010;**10**(6):817-829

[243] Ryan BM, Robles AI, Harris CC. Genetic variation in microRNA networks: The implications for cancer research. Nature Reviews Cancer. 2010;**10**(6):389-402

[244] Skeeles LE et al. The impact of 3'UTR variants on differential expression of candidate cancer susceptibility genes. PLoS One. 2013;**8**(3):e58609

[245] Ledda B et al. Small RNAs in eucaryotes: New clues for amplifying microRNA benefits. Cell & Bioscience. 2020;**10**:1

[246] Casarotto M et al. Beyond MicroRNAs: Emerging role of other non-coding RNAs in HPV-driven cancers. Cancers (Basel). 2020;**12**(5)

[247] Tjaden B et al. Target prediction for small, noncoding RNAs in bacteria. Nucleic Acids Research. 2006;**34**(9):2791-2802

[248] Lee HE, Huh JW, Kim HS. Bioinformatics analysis of evolution and human disease related transposable element-derived microRNAs. Life (Basel). 2020;**10**(6)

[249] Pegler JL et al. Miniature inverted-repeat transposable elements: Small DNA transposons that have contributed to plant MICRORNA gene evolution. Plants (Basel). 2023;**12**(5)

[250] Hu X et al. Recent advances in the functional explorations of nuclear microRNAs. Frontiers in Immunology. 2023;**14**:1097491

[251] Ortiz-Quintero B. Extracellular MicroRNAs as intercellular mediators and noninvasive biomarkers of cancer. Cancers (Basel). 2020;**12**(11)

[252] Jimenez-Avalos JA, Fernandez-MaciasJC,Gonzalez-PalomoAK. Circulating exosomal MicroRNAs: New non-invasive biomarkers of non-communicable disease. Molecular Biology Reports. 2021;**48**(1):961-967

[253] Vu LT et al. microRNA exchange via extracellular vesicles in cancer. Cell Proliferation. 2020;**53**(11):e12877

[254] Li J, Jiang X, Wang K. Exosomal miRNA: an alternative mediator of cell-to-cell communication. ExRNA. 2019;**1**(1):31

[255] Li Z, Xu R, Li N. Correction to: MicroRNAs from plants to animals, do they define a new messenger for communication? Nutrition & Metabolism (London). 2018;**15**:74

[256] Cai X, Hagedorn CH, Cullen BR. Human microRNAs are processed from capped, polyadenylated transcripts that can also function as mRNAs. RNA. 2004;**10**(12):1957-1966

[257] Lee Y et al. MicroRNA genes are transcribed by RNA polymerase II. The EMBO Journal. 2004;**23**(20):4051-4060

[258] Lee Y et al. MicroRNA maturation: Stepwise processing and subcellular localization. The EMBO Journal. 2002;**21**(17):4663-4670

[259] Singh G, Storey KB. MicroRNA cues from nature: A roadmap to decipher and combat challenges in human health and disease? Cell. 2021;**10**(12)

[260] Vilimova M, Pfeffer S. Post-transcriptional regulation of

polycistronic microRNAs. Wiley Interdisciplinary Reviews. RNA. 2023;**14**(2):e1749

[261] Bartel DP. Metazoan MicroRNAs. Cell. 2018;**173**(1):20-51

[262] Annese T et al. microRNAs biogenesis, functions and role in tumor angiogenesis. Frontiers in Oncology. 2020;**10**:581007

[263] Ha M, Kim VN. Regulation of microRNA biogenesis. Nature Reviews Molecular Cell Biology. 2014;**15**:509

[264] Abdelfattah AM, Park C, Choi MY. Update on non-canonical microRNAs. Biomolecular Concepts. 2014;**5**(4):275-287

[265] De Paolis V et al. Epitranscriptomics: A new layer of microRNA regulation in cancer. Cancers (Basel). 2021;**13**(13)

[266] Gregorova J, Vychytilova-Faltejskova P, Sevcikova S. Epigenetic regulation of MicroRNA clusters and families during tumor development. Cancers (Basel). 2021;**13**(6)

[267] Pifkova L, Vesela P, Slaby O. The development and significance of microRNA sequence variants in carcinogenesis. Klinická Onkologie. 2021;**34**(1):20-25

[268] Orban TI. One locus, several functional RNAs-emerging roles of the mechanisms responsible for the sequence variability of microRNAs. Biologia Futura. 2023. DOI: 10.1007/s42977-023-00154-7

[269] Wu D et al. Alternative splicing and MicroRNA: Epigenetic mystique in male reproduction. RNA Biology. 2022;**19**(1):162-175

[270] Zhang B, Wang Q, Pan X. MicroRNAs and their regulatory roles in animals and plants. Journal of Cellular Physiology. 2007;**210**(2):279-289

[271] Naeli P et al. The intricate balance between microRNA-induced mRNA decay and translational repression. The FEBS Journal. 2023;**290**(10):2508-2524

[272] Henke JI et al. microRNA-122 stimulates translation of hepatitis C virus RNA. The EMBO Journal. 2008;**27**(24):3300-3310

[273] Jopling CL et al. Modulation of hepatitis C virus RNA abundance by a liver-specific MicroRNA. Science. 2005;**309**(5740):1577-1581

[274] Li HC, Yang CH, Lo SY. Roles of microRNAs in hepatitis C virus replication and pathogenesis. Viruses. 2022;**14**(8)

[275] Saraiya AA, Li W, Wang CC. Transition of a microRNA from repressing to activating translation depending on the extent of base pairing with the target. PLoS One. 2013;**8**(2):e55672

[276] Vasudevan S, Tong Y, Steitz JA. Switching from repression to activation: microRNAs can up-regulate translation. Science. 2007;**318**(5858):1931-1934

[277] Syed SN, Brune B. MicroRNAs as emerging regulators of Signaling in the tumor microenvironment. Cancers (Basel). 2020;**12**(4)

[278] Hon LS, Zhang Z. The roles of binding site arrangement and combinatorial targeting in microRNA repression of gene expression. Genome Biology. 2007;**8**(8):R166

[279] Reinhart BJ et al. MicroRNAs in plants. Genes & Development. 2002;**16**(13):1616-1626

[280] Duan Y, Veksler-Lublinsky I, Ambros V. Critical contribution of 3′ non-seed base pairing to the in vivo function of the evolutionarily conserved let-7a microRNA. Cell Reports. 2022;**39**(4):110745

[281] Han J, Mendell JT. MicroRNA turnover: A tale of tailing, trimming, and targets. Trends in Biochemical Sciences. 2023;**48**(1):26-39

[282] Grimson A et al. MicroRNA targeting specificity in mammals: Determinants beyond seed pairing. Molecular Cell. 2007;**27**(1):91-105

[283] Shin C et al. Expanding the microRNA targeting code: Functional sites with centered pairing. Molecular Cell. 2010;**38**(6):789-802

[284] Garcia DM et al. Weak seed-pairing stability and high target-site abundance decrease the proficiency of lsy-6 and other microRNAs. Nature Structural & Molecular Biology. 2011;**18**(10):1139-1146

[285] Reczko M et al. Functional microRNA targets in protein coding sequences. Bioinformatics. 2012;**28**(6):771-776

[286] Zhou X et al. Abundant conserved microRNA target sites in the 5′-untranslated region and coding sequence. Genetica. 2009;**137**(2):159-164

[287] Jorgensen BG, Ro S. MicroRNAs and 'Sponging' competitive endogenous RNAs dysregulated in colorectal cancer: Potential as noninvasive biomarkers and therapeutic targets. International Journal of Molecular Sciences. 2022;**23**(4)

[288] Zhou Q et al. Novel insights into MALAT1 function as a MicroRNA sponge in NSCLC. Frontiers in Oncology. 2021;**11**:758653

[289] Cantile M et al. Functional interaction among lncRNA HOTAIR and MicroRNAs in cancer and other human diseases. Cancers (Basel). 2021;**13**(3)

[290] Nam JW et al. Global analyses of the effect of different cellular contexts on microRNA targeting. Molecular Cell. 2014;**53**(6):1031-1043

[291] Gu S et al. Functional SNP in 3'-UTR MicroRNA-binding site of ZNF350 confers risk for age-related cataract. Human Mutation. 2016;**37**(11):1223-1230

[292] Flamand MN et al. A non-canonical site reveals the cooperative mechanisms of microRNA-mediated silencing. Nucleic Acids Research. 2017;**45**(12):7212-7225

[293] Li L et al. Widespread microRNA degradation elements in target mRNAs can assist the encoded proteins. Genes & Development. 2021;**35**(23-24):1595-1609

[294] Sethupathy P et al. Human microRNA-155 on chromosome 21 differentially interacts with its polymorphic target in the AGTR1 3′ untranslated region: A mechanism for functional single-nucleotide polymorphisms related to phenotypes. American Journal of Human Genetics. 2007;**81**(2):405-413

[295] Genetic home reference. Retrieved from: http://ghr.nlm.nih.gov/handbook/genomicresearch/snp.

[296] Gartner JJ et al. Whole-genome sequencing identifies a recurrent functional synonymous mutation in melanoma. Proceedings of the National Academy of Sciences of the United States of America. 2013;**110**(33):13481-13486

[297] Cline MS, Karchin R. Using bioinformatics to predict the functional impact of SNVs. Bioinformatics. 2011;**27**(4):441-448

[298] Hu Z et al. Genetic polymorphisms in the precursor MicroRNA flanking region and non-small cell lung cancer survival. American Journal of Respiratory and Critical Care Medicine. 2011;**183**(5):641-648

[299] Diederichs S, Haber DA. Sequence variations of microRNAs in human cancer: Alterations in predicted secondary structure do not affect processing. Cancer Research. 2006;**66**(12):6097-6104

[300] Duan R, Pak C, Jin P. Single nucleotide polymorphism associated with mature miR-125a alters the processing of pri-miRNA. Human Molecular Genetics. 2007;**16**(9):1124-1131

[301] Sun G et al. SNPs in human miRNA genes affect biogenesis and function. RNA. 2009;**15**(9):1640-1651

[302] George GP, Mittal RD. MicroRNAs: Potential biomarkers in cancer. Indian Journal of Clinical Biochemistry. 2010;**25**(1):4-14

[303] Greenberg E et al. Mutagen-specific mutation signature determines global microRNA binding. PLoS One. 2011;**6**(11):e27400

[304] Wu H et al. Alternative processing of primary microRNA transcripts by Drosha generates 5′ end variation of mature microRNA. PLoS One. 2009;**4**(10):e7566

[305] Calin GA, Croce CM. MicroRNA-cancer connection: The beginning of a new tale. Cancer Research. 2006;**66**(15):7390-7394

[306] Clop A et al. A mutation creating a potential illegitimate microRNA target site in the myostatin gene affects muscularity in sheep. Nature Genetics. 2006;**38**(7):813-818

[307] Hu Z, Bruno AE. The influence of 3'UTRs on MicroRNA function inferred from human SNP data. Comparative and Functional Genomics. 2011;**2011**:910769

[308] Georges M, Coppieters W, Charlier C. Polymorphic miRNA-mediated gene regulation: Contribution to phenotypic variation and disease. Current Opinion in Genetics & Development. 2007;**17**(3):166-176

[309] Borel C, Antonarakis SE. Functional genetic variation of human miRNAs and phenotypic consequences. Mammalian Genome. 2008;**19**(7-8):503-509

[310] Lu M et al. An analysis of human microRNA and disease associations. PLoS One. 2008;**3**(10):e3420

[311] Zhang R, Su B. MicroRNA regulation and the variability of human cortical gene expression. Nucleic Acids Research. 2008;**36**(14):4621-4628

[312] Mencia A et al. Mutations in the seed region of human miR-96 are responsible for nonsyndromic progressive hearing loss. Nature Genetics. 2009;**41**(5):609-613

[313] Harnprasopwat R et al. Alteration of processing induced by a single nucleotide polymorphism in pri-miR-126. Biochemical and Biophysical Research Communications. 2010;**399**(2):117-122

[314] Slezak-Prochazka I et al. MicroRNAs, macrocontrol: Regulation of miRNA processing. RNA. 2010;**16**(6):1087-1095

[315] Saunders MA, Liang H, Li WH. Human polymorphism at microRNAs and microRNA target sites. Proceedings of the National Academy of Sciences of the United States of America. 2007;**104**(9):3300-3305

[316] Gong J et al. Genome-wide identification of SNPs in microRNA genes and the SNP effects on microRNA target binding and biogenesis. Human Mutation. 2012;**33**(1):254-263

[317] Marceca GP et al. Detecting and characterizing A-to-I microRNA editing in cancer. Cancers (Basel). 2021;**13**(7)

[318] Ma R et al. MicroRNA polymorphism: A target for diagnosis and prognosis of hepatocellular carcinoma? Oncology Letters. 2021;**21**(4):324

[319] Li Y et al. The associations and roles of microRNA single-nucleotide polymorphisms in cervical cancer. International Journal of Medical Sciences. 2021;**18**(11):2347-2354

[320] Zhou C et al. Association of miR-146a, miR-149 and miR-196a2 polymorphisms with neuroblastoma risk in eastern Chinese population: A three-center case-control study. Bioscience Reports. 2019;**39**(6)

[321] Wang K et al. A functional variation in pre-microRNA-196a is associated with susceptibility of esophageal squamous cell carcinoma risk in Chinese Han. Biomarkers. 2010;**15**(7):614-618

[322] Sun G, Li H, Rossi JJ. Sequence context outside the target region influences the effectiveness of miR-223 target sites in the RhoB 3'UTR. Nucleic Acids Research. 2010;**38**(1):239-252

[323] Kertesz M et al. The role of site accessibility in microRNA target recognition. Nature Genetics. 2007;**39**(10):1278-1284

[324] Budmark A et al. Single nucleotide variations within and around microRNA-binding sites. Cancer Research. 2018;**78**(13):490-490

[325] Berardi C et al. Multi-ethnic analysis reveals soluble L-selectin may be post-transcriptionally regulated by 3'UTR polymorphism: The multi-ethnic study of atherosclerosis (MESA). Human Genetics. 2015;**134**(4):393-403

[326] Gibson G. Rare and common variants: Twenty arguments. Nature Reviews Genetics. 2012;**13**(2):135-145

[327] Hermann T, Westhof E. Non-Watson-crick base pairs in RNA-protein recognition. Chemistry & Biology. 1999;**6**(12):R335-R343

[328] Smalheiser NR, Torvik VI. Complications in mammalian microRNA target prediction. Methods in Molecular Biology. 2006;**342**:115-127

[329] Kuc C et al. Rainbow trout exposed to benzo[a]pyrene yields conserved microRNA binding sites in DNA methyltransferases across 500 million years of evolution. Scientific Reports. 2017;**7**(1):16843

[330] Jopling CL, Schutz S, Sarnow P. Position-dependent function for a tandem microRNA miR-122-binding site located in the hepatitis C virus RNA genome. Cell Host & Microbe. 2008;**4**(1):77-85

[331] Ivashchenko A et al. The properties of binding sites of miR-619-5p, miR-5095, miR-5096, and miR-5585-3p in the mRNAs of human genes. BioMed Research International. 2014;**2014**:720715

[332] Koscianska E et al. Cooperation meets competition in microRNA-mediated DMPK transcript regulation. Nucleic Acids Research. 2015;**43**(19):9500-9518

[333] Thibault PA et al. Regulation of hepatitis C virus genome replication by Xrn1 and MicroRNA-122 binding to